Adventures in Recreational Mathematics

Volume 2

Problem Solving in Mathematics and Beyond

Print ISSN: 2591-7234
Online ISSN: 2591-7242

Series Editor: Dr. Alfred S. Posamentier
Distinguished Lecturer
New York City College of Technology - City University of New York

There are countless applications that would be considered problem solving in mathematics and beyond. One could even argue that most of mathematics in one way or another involves solving problems. However, this series is intended to be of interest to the general audience with the sole purpose of demonstrating the power and beauty of mathematics through clever problem-solving experiences.

Each of the books will be aimed at the general audience, which implies that the writing level will be such that it will not engulfed in technical language — rather the language will be simple everyday language so that the focus can remain on the content and not be distracted by unnecessarily sophiscated language. Again, the primary purpose of this series is to approach the topic of mathematics problem-solving in a most appealing and attractive way in order to win more of the general public to appreciate his most important subject rather than to fear it. At the same time we expect that professionals in the scientific community will also find these books attractive, as they will provide many entertaining surprises for the unsuspecting reader.

Published

Vol. 21   *Adventures in Recreational Mathematics*
          (In 2 Volumes)
          by David Singmaster

Vol. 20   *X Games: Training in Sports to Play in Mathematics*
          by Tim Chartier

Vol. 19   *Mathematical Nuggets from Austria: Selected Problems from the
          Styrian Mid-Secondary School Mathematics Competitions*
          by Robert Geretschläger and Gottfried Perz

Vol. 18   *Mathematics Entertainment for the Millions*
          by Alfred S Posamentier

For the complete list of volumes in this series, please visit www.worldscientific.com/series/psmb

Problem Solving in
Mathematics and Beyond | Volume **21**

Adventures in Recreational Mathematics

Volume 2

David Singmaster
London South Bank University, UK

World Scientific

NEW JERSEY · LONDON · SINGAPORE · BEIJING · SHANGHAI · HONG KONG · TAIPEI · CHENNAI · TOKYO

Published by

World Scientific Publishing Co. Pte. Ltd.

5 Toh Tuck Link, Singapore 596224

USA office: 27 Warren Street, Suite 401-402, Hackensack, NJ 07601

UK office: 57 Shelton Street, Covent Garden, London WC2H 9HE

Library of Congress Cataloging-in-Publication Data

Names: Singmaster, David, author.
Title: Adventures in recreational mathematics / David Singmaster, London South Bank University, UK.
Description: Hackensack, NJ : World Scientific, [2022] | Series: Problem solving in mathematics
    and beyond, 2591-7234 ; vol. 21 | Includes bibliographical references and index.
Identifiers: LCCN 2020037888 (print) | LCCN 2020037889 (ebook) |
    ISBN 9789811225642 (set ; hardcover) | ISBN 9789811226304 (set ; paperback) |
    ISBN 9789811226007 (vol. 1 ; hardcover) | ISBN 9789811226502 (vol. 1 ; paperback) |
    ISBN 9789811226038 (vol. 2 ; hardcover) | ISBN 9789811226519 (vol. 2 ; paperback) |
    ISBN 9789811226014 (vol. 1 ; ebook for institutions) |
    ISBN 9789811226021 (vol. 1 ; ebook for individuals) |
    ISBN 9789811226045 (vol. 2 ; ebook for institutions) |
    ISBN 9789811226052 (vol. 2 ; ebook for individuals)
Subjects: LCSH: Mathematical recreations. | Mathematical recreations--History.
Classification: LCC QA95 .S4958 2022 (print) | LCC QA95 (ebook) | DDC 793.74--dc23
LC record available at https://lccn.loc.gov/2020037888
LC ebook record available at https://lccn.loc.gov/2020037889

British Library Cataloguing-in-Publication Data
A catalogue record for this book is available from the British Library.

For any available supplementary material, please visit
https://www.worldscientific.com/worldscibooks/10.1142/11977#t=suppl

Desk Editors: Vishnu Mohan/Tan Rok Ting

Typeset by Stallion Press
Email: enquiries@stallionpress.com

Printed in Singapore

Dedicated to the memory of great puzzlers and friends:
Ray Bathke, John Conway, Martin Gardner, Richard Guy,
Solomon Golomb, Edward Hordern and Nob Yoshigahara.

Contents of Volume 2

Preface

My first published paper was "On round pegs in square holes and square pegs in round holes" (1964), reproduced as Chapter 2. Martin Gardner mentioned the result in his *Scientific American* column. This encouraged me to not only persist with this problem, but to engage in other recreational topics.

This interest became serious after the Rubik's Cube craze in 1978 and onward. I then thought it would be possible to produce a book giving the origins of recreational problems. When I embarked on this, I soon discovered that this information had never been collected and was largely unknown. This led to decades of research into the origins of these puzzles and problem. This is the subject of the companion volume *Adventures in Recreational Mathematics*.

However, even fairly recent puzzles have obscure origins. For example, the famous problem where one has 12 coins, one of which is counterfeit and weighs more or less than the others and one has to find which coin is counterfeit and whether it is heavy or light in three weighings appears to have evolved within living memory, during World War II, but no one claims to have invented it. This volume is devoted to recently posed problems.

The opening chapter explains the breadth of the topic, and I use a wide definition. Others might use a narrower scope. When dealing with recently posed problems every mathematical discussion uses mathematical tools. Those tools have a history and that history may lead back to a recreational topic. In this volume, we usually do not follow such threads back in time, focusing instead on the new problem.

The mathematical sophistication varies between chapters. Some of the articles collected in this volume appeared in undergraduate mathematics journals. There is no advanced mathematics herein, but sometimes I assume the reader is a beginning undergraduate. In general, this volume requires more knowledge than what was required for the companion volume. This means if you only have mastered high school algebra and geometry there may be some passages that you will want to skim over. When prior experience is lacking, you are only asked to accept the prior knowledge quoted. Hopefully any passages that are not clear on first reading will encourage further study.

The first chapter is more discursive, discussing the utility of recreational mathematics. In doing so it briefly goes over some historical facts covered in more detail in the companion volume. In each subsequent chapter I pose a problem. In some cases it continues some recent research by others. In other cases it is a problem that I may be the first to pose. However, generally, the rest of this book is a series of chapters motivated by curiosity.

Mathematics is the "queen of the sciences". And like science it is a combination of curiosity-based research and settling old conjectures. This book is mainly curiosity-based. What makes it "recreational" is the very fact that the problems excite our interest without the need for some application to motivate them. We have fun solving them because we are curious what the answer is. This is in contrast to restricting ourselves to "puzzles".

For example, I studied the Sum = Product sequences for some time in the late 1980s. This concerns sets of numbers, like $\{1, 2, 3\}$ which sum to the same number as their product. There is no need for an application to pursue this topic, we just want to know. Further, any setter of puzzles worth his or her salt can take the results (in Chapter 4) and make a puzzle out of them.

The final chapter is more of a coda. It takes a recreation of uncertain age, which is more of an optical illusion than a puzzle, and shows how widespread it is today.

I hope you find these problems as entertaining as I found them when solving them.

About the Author

 **David Breyer Singmaster** studied at Caltech and received a Ph.D. in mathematics from the University of California-Berkeley in 1966. He taught at the American University of Beirut, later lived in Cyprus, and then came to London in 1970 — and has been based there since. He retired from London South Bank University in 1996 and was designated emeritus in 2020. His interests are in number theory and combinatorics, and the history of mathematics and of science in general.

From 1978 to about 1984, he was the leading expositor of Rubik's Cube. He devised the now standard notation for it, wrote the first book on the Cube, and later edited Rubik's book into English. Due to revived interest in the Cube, he and some colleagues produced a new book *The Cube: The Ultimate Guide to the World's Bestselling Puzzle — Secrets, Stories, Solutions* in 2009 and *The Handbook of Cubik Math* in 2010.

Since about 1982, he has been working on a history of recreational mathematics, *Sources in Recreational Mathematics*, see Appendix A of Vol. 1, which has involved reading and studying mathematics from every culture and period.

He was the opening speaker at the Conference opening the Strens Memorial Collection at Calgary in 1986. He has attended all of the Gatherings for Gardner and spoken at many of them. He acted as Chairman for four of these. He has attended all of the MathsJam's in England. He was an invited speaker at the Third

Iberian Colloquium on Recreational Mathematics in 2012, where he spoke on "Vanishing Area Puzzles" (Chapter 16 of Vol. 1). He attended the Fourth Colloquium in Lisbon in 2015, where he spoke on "Early Topological Puzzles" (Chapter 8 of Vol. 1). His book *Problems for Metagrobologists: A Collection of Puzzles With Real Mathematical, Logical or Scientific Content*, a collection of over 200 problems that he composed since 1988, appeared in 2015.

Chapter 1

Why Recreational Mathematics?

"Amusement is one of the fields of applied mathematics."
— William F. White, *A Scrap-Book of Elementary Mathematics*, 1908.

This chapter is inspired by Eugene Wigner's famous essay "The unreasonable effectiveness of mathematics in the physical sciences" [35]. Unlike Wigner, after much thought, we present an explanation of why recreational mathematics is, if not "reasonable", at least useful.[1] Wigner was concerned with the philosophy of science, but our concerns are pragmatic, not philosophical.

1.1 The Nature of Recreational Mathematics

To begin with, it is worth considering what is meant by recreational mathematics. An obvious definition is that it is mathematics that is fun. However, almost any mathematician enjoys her/his work. There are two more specific, somewhat overlapping, definitions that cover most of what is meant by recreational mathematics.

Recreational mathematics is mathematics that is fun and popular — that is, the problems should be understandable to the interested lay person, though the solutions may be harder. (However, if

---

[1]This chapter is an amplification of my talks at the European Congress of Mathematicians in 1992 [31] and at the University Mathematics Teaching Conference, September 1999 [32]. They were combined in the *UMAP Journal* 37(4) (2016), 339–380.

the solution is too hard, this may shift the topic away from the recreational; e.g., Fermat's Last Theorem, the Four-Color Theorem or the Mandelbrot Set.)

Recreational mathematics is mathematics that is fun and used pedagogically either as a diversion from serious mathematics or as a way of making serious mathematics understandable or palatable. These pedagogic uses of recreational mathematics are already present in the oldest known mathematics and continue to the present day.

Another educational use is to provide entertaining settings for standard problems to make the problems and solutions amusing and memorable. Consider problem 79 of the Rhind Papyrus. It leads to adding $7 + 49 + 343 + 2401$ in a context similar to the "As I was going to St. Ives" children's rhyme, which is known in several European countries. The problem is not thought to be an exercise in summing a geometric progression - it has no connection with other problems in the papyrus. It seems to be inserted solely as a diversion or recreation.

These two aspects of recreational mathematics — the popular and the pedagogic — overlap considerably, and there is no clear boundary between them and "serious" mathematics. In addition, there are several other independent fields that contain much recreational mathematics: games; mechanical puzzles; magic; art.

Games of chance and games of strategy also seem to be about as old as human civilization. The mathematics of games of chance began in the Middle Ages, and its development by Fermat and Pascal led to probability theory. The mathematics of games of strategy started only about the beginning of the 20th century, but soon developed into game theory, both of the von Neumann–Morgenstern type (economics based) and later of the Conway type (configuration based).

Mechanical puzzles range widely in mathematical content. Some only require a certain amount of dexterity and three-dimensional ability; others require ingenuity and logical thought; while others require systematic application of mathematical ideas or patterns, such as Rubik's Cube, the Chinese Rings, and the Tower of Hanoi.

The creation of beauty often leads to questions of symmetry and geometry that are studied for their own sake — e.g., the carved stone balls that we will see later. This outlines the conventional scope of recreational mathematics, but there is some variation due to personal taste.

1.2 The Utility of Recreational Mathematics

How is recreational mathematics useful?

Recreational problems often lead to serious mathematics

The most obvious fields are probability and graph theory, where popular problems have been a major (or even dominant) stimulus to the creation and evolution of the subject. Further reflection shows that number theory, topology, geometry, and algebra have all been strongly stimulated by recreational problems. (Though geometry has its origins in practical surveying, the Greeks treated it as an intellectual game; and much of their work must be considered as recreational in nature, even though they viewed it more seriously, as reflecting the nature of the world. From the time of the Babylonians, algebraists tried to solve cubic equations, though they had no practical problems that led to cubics.) There are even recreational aspects of calculus — e.g., the many curves studied since the 16th century. Consequently, the study of recreational topics is necessary to understanding the history of many, perhaps most, topics in mathematics. Before Aristotle, the Greeks used logic as a game of forcing an opponent to accept your conclusions, but had never formalized the rules. Aristotle began the study of logic in order to formalize the rules of this game.

Recreational mathematics often turns up ideas of unforeseen utility

A few examples will be mentioned later. Such unusual developments, and the more straightforward developments of the previous point, demonstrate the historical principle of "The (unreasonable) utility of recreational mathematics". This and similar ideas are the historical and social justification of mathematical research in general and for the study of recreational mathematics. We return to this point later.

Recreational mathematics has great pedagogic utility

This will be the main theme of our examples.

Recreational mathematics is very useful to the historian of mathematics

Recreational problems often are of great age and usually can be clearly recognized; they serve as useful historical markers, tracing the development and transmission of mathematics (and culture in general) in place and time. The Chinese Remainder Theorem, Magic Squares, Cistern Problems, and the Hundred Fowls Problem are excellent examples of this process.

> The original Hundred Fowls Problem, from 5th century China, has a man buying 100 fowls for 100 cash (an old coin). Roosters cost 5, hens 3, and chicks are 3 for a cash — how many of each did he buy?

The number of topics that have their origins in China or India is surprising and emphasizes our increasing realization that modern algebra and arithmetic derive more from Babylonia, China, India, and the Arabs than from Greece.

1.3   Some Examples of Useful Recreational Mathematics

Examples are outlined to show how recreational mathematics has been useful, perhaps stretching "recreational" a bit to include some other non-practical topics.

From gambling bets to the insurance industry

The theory of probability and statistics grew from the analysis of gambling bets to the basis of the insurance industry in the 17th and 18th centuries. Much of combinatorics likewise has its roots in gambling problems. The theory of Latin squares began as a recreation but has become an important technique in experimental design (and was reclaimed in connection with Sudoku puzzles).

From Euclid to the Moon and to Buckyballs

Greek geometry, though it had some basis in surveying, was largely an intellectual exercise, pursued for its own sake. The conic sections were developed with no purpose in mind, but 2000 years later turned out to be just what Kepler and Newton needed and were what took men to the Moon. The regular, quasi-regular, and Archimedean polyhedra were developed long before they became the basis of molecular structures. Indeed, the regular solids are now known to be prehistoric. Beginning in 1985, chemists became excited about fullerenes, molecules of carbon in various polyhedral shapes, of which the archetype is the truncated icosahedron, with 60 carbon atoms at the vertices, named *buckminsterfullerene* after Buckminster Fuller (1895–1983), of geodesic dome fame. Spherical fullerenes are consequently nicknamed "buckyballs" (see Figure 1.1). Such molecules apparently are the basis for the formation of soot particles in the air. The idea of making such molecules seems to have originated with David Jones, the scientific humorist who writes as "Daedalus," in one of his humor columns. Chemists have also synthesized hydrocarbons in the shapes of a cube (cubane, C_8H_8, in 1964) and a dodecahedron (dodecahedrane, $C_{20}H_{20}$, in 1982).

In a similar vein, non-Euclidean geometry was developed long before Einstein considered it as a possible geometry for space.

From river-crossing puzzles to graph theory

The river-crossing problems and the problem of getting camels across a desert, which occur in Alcuin, c. 800 [1], are considered to be the earliest combinatorial optimization problems. Such problems are now solved by graph-theoretic methods, dynamic programming, or integer programming. The problems of the Bridges of Königsberg, mazes, knight's tours, and circuits on the dodecahedron (the Icosian Game) (Figures 1.2 and 1.3) were major sources of graph theory and are the basis of major fields of optimization, leading to one of the major unsolved problems of the 20th century: Does P = NP? The routes of postmen, street-sweepers and snowplows, as well as those

Figure 1.1.   Frame of a truncated icosahedron buckyball, C60, illustrated in the Ambrosiana manuscript of Luca Pacioli's De divina proportione (1509) by Leonardo da Vinci. Plate XXI–III, folio 103 recto.
Source: Veneranda Biblioteca Ambrosiana, DeAgostini Picture Library, Scala, Florence.

of salesmen, are worked out by these methods. Further, Hamilton's thoughts on the Icosian Game led him to the first presentation of a group by generators and relations [19].

From number theory to splicing phone cables

Number theory is another of the fields where recreations have been a major source of problems. Further, these problems have been a major source for modern algebra. Fermat's Last Theorem led to Kummer's invention of ideals and most of algebraic number theory. There was a famous application of primitive roots to the splicing of telephone cables to minimize interference and crosstalk [28] (pp. 397–399). Primality and factorization were traditionally innocuous recreational

THE ICOSIAN GAME.

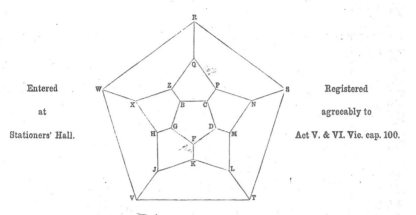

IN this new Game (invented by Sir WILLIAM ROWAN HAMILTON, LL.D., &c., of Dublin, and by him named *Icosian*, from a Greek word signifying "twenty") a player is to place the whole or part of a set of twenty numbered pieces or men upon the points or in the holes of a board, represented by the diagram above drawn, in such a manner as always to proceed *along the lines* of the figure, and also to fulfil certain *other* conditions, which may in various ways be assigned by another player. Ingenuity and skill may thus be exercised in *proposing* as well as in *resolving* problems of the game. For example, the first of the two players may place the first five pieces in any five consecutive holes, and then require the second player to place the remaining fifteen men consecutively in such a manner that the succession may be *cyclical*, that is, so that No. 20 may be adjacent to No. 1; and it is always possible to answer any question of this kind. Thus, if B C D F G be the five given initial points, it is allowed to complete the succession by following the alphabetical order of the twenty consonants, as suggested by the diagram itself; but after placing the piece No. 6 in the hole H, as before, it is *also* allowed (by the supposed conditions) to put No. 7 in X instead of J, and then to conclude with the succession, W R S T V J K L M N P Q Z. Other Examples of Icosian Problems, with solutions of some of them, will be found in the following page.

LONDON:

PUBLISHED AND SOLD WHOLESALE BY JOHN JAQUES AND SON, 102 HATTON GARDEN;
AND TO BE HAD AT MOST OF THE LEADING FANCY REPOSITORIES
THROUGHOUT THE KINGDOM.

Figure 1.2. Advertisement with instructions for Hamilton's Icosian Game. Only four original boards of the plane version are known to exist.

pastimes; but since 1978 when Rivest, Shamir, and Adleman introduced their method of public-key cryptography (now known as RSA cryptography), the factorization of a big number and the determination of the next Mersenne prime are generally front-page news.

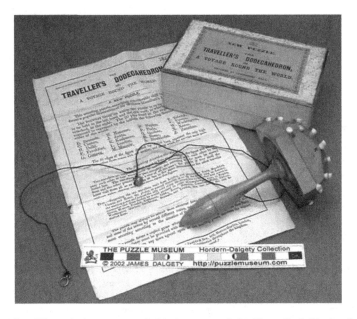

Figure 1.3. The only known remaining instance of the Traveller's Dodecahedron, a revision by Hamilton of his Icosian game with simpler rules. The 30 edges on the head represent roads to use to visit 20 ivory pegs that represent cities. "Two travellers set off visiting four neighboring towns. One returns home and the other continues to travel around the world trying to visit all the remaining cities once only". Photo courtesy of James Dalgety.

From buying a horse to negative numbers

A major impetus for algebra has been the solving of equations. The Babylonians already gave quadratic problems where the area of a rectangle was added to the difference between the length and the width. This clearly had no practical significance. Similar impractical problems led to cubic equations and the eventual solution of the cubic. Negative solutions first become common in medieval puzzle problems about men buying a horse or finding a purse.

From knots to DNA

Even in analysis, the study of curves (e.g., the cycloid) had some recreational motivation.

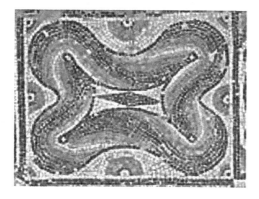

Figure 1.4. Möbius strip with five half-turns in a Roman floor mosaic of Orpheus charming the animals, ca. 200 A.D., now in the Museum of Pagan Art, Arles, France.

Topology has much of its origins in recreational aspects of curves and surfaces. Knots, another field once generally considered of no possible use, are now of great interest to molecular biologists who have discovered that DNA molecules form into closed chains which may or may not be knotted.

The Möbius strip arose about 1858 in work by both August Ferdinand Möbius (1790–1868) and Johann Benedict Listing (1808–1882), Listing being apparently a bit earlier. Depictions of it occur in Roman mosaics as noticed by Charles Seife in 2002, including a strip with five half-turns, see Figure 1.4 [25]. There are a number of other practical uses for the Möbius strip, such as conveyor belts that wear both sides evenly, but the most unusual is as a non-inductive electrical resistor – Martin Gardner said it has been patented [9], see Figure 1.5. Those who still have dot-matrix printers may (or may not) know that their printer ribbons commonly have a twist (so they are Möbius strips!) in order to allow the printer to use both edges.[2]

---

[2]I first discovered this when I found one of my department technicians trying to put such a ribbon back into its case; he had done it several times, but it kept coming out twisted, which he thought was his mistake!

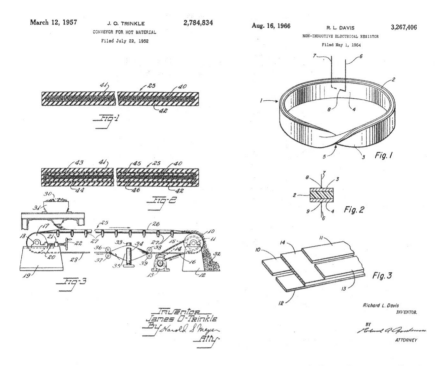

Figure 1.5.　Images from patents: Single-sided conveyor belt and a non-inductive electrical resistor.

From Chinese rings to binary codes

The Chinese Rings puzzle (Figure 1.6), known as *baguenaudier* ("time-waster") in French, may indeed have originated in China 1800 years ago. In combinatorics, the solution pattern of the Chinese Rings puzzle is the binary coding known as the Gray Code, patented as an error-minimizing code by Frank Gray (1887–1969) of Bell Labs in 1953 (Figure 1.6) and already used in the same way by Émile Baudot (1845–1903) in the 1870s [4, 5] in inventing the predecessor of the teletype. (It is from Baudot's work that the term "baud" arose as a measure of transmission speed.)

A very different code used by Baudot has a long history in India. Sometimes called a Chain Code, it is said to have been used by Sanskrit poets in about 1000 BCE (and perhaps much earlier) to memorize all the combinations of long and short syllables [34]. The

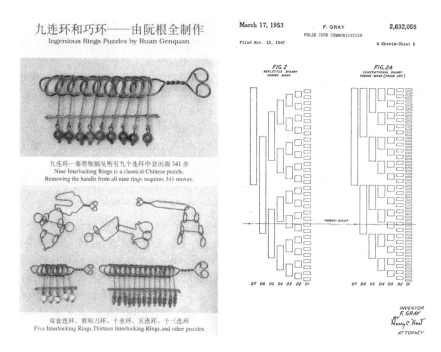

Figure 1.6. Left: Chinese Rings puzzles. The task is to remove all the rings [36]. Right: Figures from Gray's patent for pulse code communication.

10 syllables in the Sanskrit nonsense word

ya-mā-tā-rā-ja-bhā-na-sa-la-gām

contain in successive groups of three all the triplets of long beats (marked with a bar over the a) and short beats (unmarked a). Moreover, since the last two syllables are the same form as the first two, if we regard the sequence of syllables as wrapping around, we could arrange the syllables in the form of a wheel.

Baudot redesigned his printing telegraph to use a chain code [4, 20, 21]. The idea of a chain code led to the more general mathematical concept of a de Bruijn sequence [11, 18]. In a de Bruijn sequence (also known as a shift-register sequence), every possible subsequence of a prescribed length from an alphabet of characters appears exactly once in the sequence, which like a memory wheel cycles back on itself. For example, the de Bruijn sequence

0 0 0 1 0 1 1 1

contains in order all the different subsequences of length 3:

000, 001, 010, 101, 011, 111, 110, 100.

Such codes are painted on factory and warehouse floors to enable robots to determine where they are by scanning a small section of the floor. They have also been used as the basis of card tricks — divinations — where the values of cards are determined from a small amount of information [11]. Diaconis and Graham note some confusion of Gray codes and chain codes:

> [Magicians] mistakenly call de Bruijn sequences "Gray codes".
> ... But as far as we know, there has never been a single use [of
> Gray codes] in magic [11] (p. 25).

The earliest mention of what later became known as chain codes and de Bruijn sequences seems to be by Flye Sainte-Marie [13] though Baudot's use of it for a teleprinter dates from about 1882, with equipment using it exhibited in 1889 [20, 21]. Kerr [24] offers some brief mechanical details of a production teletype machine that used a chain code.

1.4   Recreational Mathematics with Objects

Neolithic polyhedra

These "carved stone balls" date from c. 2500 BCE, and occur in eastern Scotland. Examples are in the Royal Scottish, Ashmolean, Dundee, and Aberdeen Museums. Figure 1.7 shows a resin model of a carved stone ball from Aberdeenshire, made by an artist in Glastonbury. No one knows the purpose of these, but shapes in the form of all the platonic solids appear.

Plimpton 322 tablet

This is the famous Old Babylonian tablet, c. 1800 BCE, listing Pythagorean triples, as shown in Figure 1.8.[3]

---

[3]Some years ago I persuaded Columbia University to make casts from the original, and Figure 1.8 is a photograph of one of those. I believe copies may still be available from the Rare Book Room, Columbia University.

Figure 1.7. A resin model of a neolithic carved stone ball (about 9 cm across). Photo by David Singmaster.

Figure 1.8. Facsimile of the Old Babylonian tablet Plimpton 322 that lists Pythagorean triples. Photo by David Singmaster [27].

Roman dodecahedra

Approximately 100 of these are known, from Roman sites north of the Alps. The one shown in Figure 1.9 was found in 1939 in Tongeren, Belgium, and dates to 150–400 CE [22]. Its total height is 81 mm; the height without the balls at the corners is 66 mm. There is a

Figure 1.9. Roman dodecahedron found at Tongeren, Belgium, in 1939 and now situated in the Gallo-Roman Museum Tongeren. Reproduced full size in the *UMAP Journal* 4(2016). Photo by Guido Schalenbourg, Gallo-Roman Museum Tongeren.

somewhat smaller example at the Hunt Museum in Limerick. No one knows their purpose [17]. Nevertheless, they admirably match the description of Hamilton's Traveller's Dodecahedron of Figure 1.3.

Chinese magic square

The cast-iron facsimile in Figure 1.10 (cast at reduced size) is one of the five cast-iron examples of a 6×6 magic square excavated near Xi'an, China, in 1956. It is inscribed in East Arabic numerals (similar to those still used in the Middle East) and dates to the Yuan Dynasty (1280–1368) [26].

1.5 Examples of Medieval Problems

Fibonacci numbers

The Fibonacci numbers were known to ancient Sanskrit poets, from an uncertain date about 2000 years ago. The number of different patterns of fixed length of long syllables and short syllables, where

Figure 1.10. Cast-iron facsimile of a Chinese magic square dating to the Yuan Dynasty (1280–1368). A present from Jerry Slocum, who got it in Xi'an. Photo by David Singmaster.

a long syllable is twice as long as a short syllable, is a Fibonacci number. For example, the patterns with total length the equivalent of four short syllables are LL, SSL, SLS, LSS, and SSSS, for a total of five. However, the first Indian work in which mathematical investigation was made of such numbers was not until 1356, where they were related to binomial coefficients [29].

The Josephus problem

This is the problem of recursively counting out every kth person, and removing those, from a circle of n people. Early versions counted out half the group (see Figure 1.12 [7]); later authors and the Japanese are interested in the last man — the survivor. Euler (1775) seems to be the first to ask for the last man in general. Cardan (1539) [8] is the first to associate this process with Josephus [23]:

> A company of 40 soldiers, along with Josephus himself, were trapped in a cave by Roman soldiers during the Siege of Yodfat in 67 AD. The Jewish soldiers chose to die rather than surrender, so they devised a system to kill off each other until only one person remained. (That last person would be the only one required to die by their own hand.)

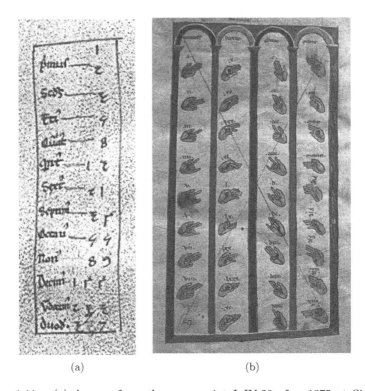

(a)　　　　　　　　　　　　　　(b)

Figure 1.11.　(a) A page from the manuscript L.IV.20 of c. 1275 at Siena of Fibonacci's *Liber abbaci* of 1202 and 1228 [12]. This manuscript, which also includes his hand signs for numbers (image (b)), is apparently the earliest known extant version of his book. The page shows the Fibonacci sequence $1, 2, 3, 5, \ldots, 377$, where each entry is the sum of the two preceding. Fibonacci introduced the sequence in a fanciful model for the number of rabbit pairs in successive generations. Photo by David Singmaster.

The original account said they formed a circle and each, in turn, killed the soldier to the left. The more popular telling is that every third soldier was killed. Some later authors derive this from the Roman practice of decimation.

River-crossing problems

The *Propositiones ad acuendos juvenes*, attributed to Alcuin of York, c. 800, contains classic river-crossing problems such as the *wolf, goat, and cabbage* problem.

Figure 1.12. Calandri's version of the Josephus problem, with 15 Franciscan (in brown) and 15 Camoldensian (in white) monks on a boat, and counted out by $k = 9$. Where should the standing monk start counting by 9, and in which direction, so that all the white-robed monks are counted out?

> A farmer must transport a wolf, a goat, and a cabbage across the river in a boat that can hold only the farmer and one other item; the restrictions are that the wolf cannot be left on either bank with the goat, nor the goat with the cabbage, unaccompanied by the farmer. (See Figure 1.13.)

The problems above are among the earliest combinatorial optimization problems. Martin Grötschel in Berlin uses the wolf–goat–cabbage problem to teach integer programming; his class found a shorter solution, but it involved halving the cabbage and halving the wolf! [6].

1.6 Examples of Modern Recreational Problems

Longest fishing pole in a box

A familiar problem involves a fisherman (or hunter or skier) who wants to mail his 2.5 m fishing pole (or gun or skis) and finds that

Figure 1.13.   Wolf–goat–grain puzzle from Columbia Algorism [10].

the post office has a maximum parcel length of 1.5 m. The fisherman solves the problem by making a cubical box of edge 1.5 m and putting the pole in diagonally. The diagonal of the box is $1.5 \times \sqrt{3} \approx 2.598$. This is very ingenious, but unfortunately there are other postal regulations. The length plus the girth must be at most 3 m. The girth is the circumference in a plane perpendicular to the longest dimension, which is the length. For a box of dimensions $A \times B \times C$, with $A \geq B \geq C$, the girth is $2B + 2C$; and so we must have $A \leq 1.5$ and $A + 2B + 2C \leq 3$. What is the longest fishing pole that can be posted under these limitations? The problem of finding the largest *volume* that one can mail with girth constraints is well known, and the maximum occurs for a cylindrical tube.

Crossing a field

The following seems as if it should be an easy question.

> You are on a path which runs south to a road. Along the road is a bus stop, and you want to get to it as quickly as possible. Between the path and the road is a field; and you can cut across the field, but your speed may be slower than on the path or the road. Is it ever the case that the optimal route is to go part way along the path, then go obliquely across the field to a point part way along the road, and then go the rest of the way along the road? Try to convince yourself of the answer before doing any calculations. Determine the optimal route in general for all situations.

More formally, let us assume that you start at the point $A = (0, W)$ and are traveling to $D = (L, 0)$ and that you travel from

A to $B = (0, y)$, thence to $C = (x, 0)$, then to D. There are three speeds involved, but only their relative values are important, so you can assume that your speed on the road is a unit speed, your speed on the path is v, and your speed on the field is V, with $V \leq v \leq 1$.

There is a common feature of the problems of crossing a field and posting the longest fishing pole that you might discover when you solve them, which makes these problems more interesting.

The Penrose pieces

One of the basic results of crystallography is that no crystal structure can have five-fold symmetry. Roger Penrose had long been interested in tiling the plane with pieces that could not tile the plane periodically. He found such a tiling with six kinds of shape and then managed to reduce it to two shapes, kites and darts, see Figure 1.14, that could tile the plane in uncountably many ways but in no periodic way. Some of the tilings have a five-fold center of symmetry, and all have a sort of generalized five-fold symmetry. They are now called "quasicrystals". These tilings fascinated both geometers and crystallographers and were extensively studied since the 1970s.

Penrose's "kites and darts" shapes were simplified further to "fat and thin rhombuses" (Figure 1.14). Rules for putting them together (e.g., a side corner of a kite must coincide with the tip or rear of a dart) prevent the shapes from tiling periodically. Figure 1.15 shows a "Penrose pattern" made from the rhombuses of Figure 1.14.

The rhombus shapes were extended to three dimensions, where they are related to the rhombic triacontahedron. Though the tilings are not periodic, they have quasi-axes and quasi-planes, which

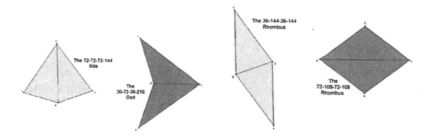

Figure 1.14. Penrose's dart and kite, and his fat and thin rhombuses, with notation of the degrees of the angles involved. Courtesy of Robert Austin.

Figure 1.15. A Penrose pattern made from the rhombuses of Figure 1.14. Courtesy of Robin Wilson.

can cause diffraction. Using these, crystallographers determined the diffraction pattern that a hypothetical quasicrystal would produce: It has a ten-fold center of symmetry.

In 1984, such diffraction patterns were discovered by Dan Shechtman (b. 1941) in a sample of rapidly cooled alloy now known as shechtmanite; and some 20 substances are now known to have quasi-crystalline forms. Shechtman received the 2011 Nobel Prize in Chemistry for his discovery of quasicrystals. Indeed, examples had been found about 30 years earlier but the diffraction patterns were discarded as being erroneous! It is not yet known whether such materials will be useful but they may be harder or stronger than other forms of the alloys and hence may find use on airplanes, rockets, etc. So a mathematical flight of fancy has led to the discovery of a new kind of matter on which we may be flying in the future! (See [15, 16] for expositions of this topic; also see the companion volume.)

1.7   The Educational Value of Recreations

Problem solving

Rubik's Cube[4] is an excellent example of problem solving (Figure 1.16 shows the bigger version that is sometimes known

---

[4]Registered trademark. I was author of one of the first books about Rubik's Cube [30] and am a co-author of others [14, 33].

Figure 1.16. A larger-than-usual (5 × 5 × 5) Rubik cube, scrambled; there are even 8 × 8 × 8, 9 × 9 × 9 and larger cubes.

as the Professor's Cube). One can identify many of the classic problem-solving skills with this example:

- understanding the problem;
- establishing a notation;
- investigating subproblems;
- using problem reductions, which is a special case of one of the basic problem-solving techniques — transform a problem to a situation one knows how to solve and then transform the answer back to the original situation — this is the idea of logarithms, Laplace, Fourier and other transforms, similarity transformations, change of basis, mathematical modeling, etc.;
- creating an algorithm;
- demonstrating correctness of the algorithm; and
- seeking an optimum algorithm.

This puzzle excited unprecedented interest in the public demonstrating the wide appeal of problem solving. Mathematicians are also interested in exploring it further. For example, in 2010, Tomas Rokicki and friends showed that it takes at most 20 moves to restore an ordinary Rubik's Cube. The world of "twisty" puzzles has exploded with wild complexity. For example, the eminent Dutch puzzle designer, Oskar van Deventer, has designed and made a 17×17×17 cube!

A treasury of problems

Recreational mathematics is a treasury of problems which make mathematics fun. These problems have been tested by generations going back to about c. 1800 BCE. In medieval arithmetic texts, recreational questions are interspersed with more straightforward problems to provide breaks in the hard slog of learning. These problems are often based on reality, though with enough whimsy so that they have appealed to students and mathematicians for years. They illustrate the idea that "Mathematics is all around you — you only have to look for it".

An optimal learning experience

"A good problem is worth a thousand exercises" (ancient proverb, made up by myself). There is no greater learning experience than trying to solve a good problem. Recreational mathematics provides many such problems and almost every problem, after it is solved, can be amended or extended.

Solving problems naturally develops problem-solving techniques. Some of those which arise in recreational problems are as follows:

- The problems often require clarification of the assumptions and one may vary the assumptions to get different problems.
- The mathematical or logical methods needed are often nonstandard and hence one has to use basic ideas in a novel way.
- The problems are often open-ended and natural generalizations are often unsolved, so one has to re-examine the problem and ask new questions.
- For better or worse, mathematics is one of the only school courses where students are expected to learn how to think! But thinking, like problem solving, is best learned by doing and our problems are ideal for encouraging this.

Communication of mathematical ideas to the public

A primary use of recreational mathematics is that it provides us a way to communicate mathematical ideas to the public at large.

Unfortunately, mathematicians tend to underestimate the public interest in mathematics. We all know the social situation when you confess that you are a mathematician and the response is, "Oh. I was never any good at maths". Yet somewhere approaching 200 million Rubik Cubes were sold in three years! Indeed, there were more Rubik Cubes sold in Hungary than there were people. In contrast, the best-selling game is Monopoly, took 50 years to sell about 90 million examples.

Another measure of the popularity of recreational mathematics is the number of books that appear in the field each year, perhaps 50 in English alone. The long-term best-selling recreational book in English must be *Mathematical Recreations and Essays* by W. W. Rouse Ball (1850–1925), originally published in 1892 and now in its 13th edition [3]. It has rarely been out of print in that time. And there are many older books, such as *Problèmes plaisants et délectables* ... by Claude Gaspard Bachet de Méziriac in 1612, which had three editions in the late 19th century, the last of which was reprinted several times in the 20th century [2]. Many newspapers and professional magazines run regular mathematical puzzles, though this was perhaps more common in the past. Henry Dudeney published weekly columns for about 15 years and then monthly columns for about 20 years. Martin Gardner's columns for 25 years were a major factor in the popularity of *Scientific American* and probably inspired more students to study mathematics than any other influence. It has been reported that circulation dropped noticeably when he retired, it has been reported. Other names of major columnists/writers promoting mathematics with a recreational tone are:

- In English:
  - "Lewis Carroll" (= Charles Lutwidge Dodgson) (1832–1898);
  - Sam Loyd (1841–1911);
  - "Professor Hoffmann" (= Angelo John Lewis) (1839–1919);
  - "Caliban" (= Hubert Phillips) (1891–1964);
  - Thomas H. O'Beirne (1915–1982);
  - Douglas St. Paul Barnard (b. 1924);
  - Henry Dudeney (1857–1930);
  - Martin Gardner (1914–2010);
  - Ian Stewart (b. 1945).
  - Chris Maslanka (b. 1954).

- In German:
  - Wilhelm Ahrens (1912–1998);

  - Hermann Schubert (1848–1911);
  - Walther Lietzmann (1880–1959);
  - Heinrich Hemme (b. 1955).

- In French:

  - Édouard Lucas (1842–1891);
  - Pierre Berloquin (b. 1939).

There really is considerable interest in mathematics out there; and if we enjoy our subject, it should be our duty and our pleasure to try to encourage and feed this interest.[5] Indeed, it may be necessary for our self-preservation.

Further, because of its long history, recreational mathematics is an ideal vehicle for communicating the historical and multicultural aspects of mathematics.

1.8   Why Is Recreational Mathematics So Useful?

My answer to this perhaps also answers Wigner's question. Mathematics has been described as a search for pattern — that certainly describes much of what we do and also much of what most scientists do. But how do we find patterns? The real world is messy and patterns are difficult to see. As we begin to see a pattern, we tend to remove all of the inessential details and get to an ideal or model situation. These models may be so removed from reality that they become fanciful — or even recreational.

For example, physicists deal with frictionless perfectly elastic particles, weightless strings, ideal gases, etc.; mathematicians deal with random samples, exact measurements, negative money, etc. Then such models get modified and adapted into a large variety of models and techniques are developed to describe and solve them.

Now, one of the ways in which a science progresses is by seeing analogies between reality and simpler situations. For example, the idea of the circulation of the blood could not be developed until the idea of a pump was known and somewhat understood. The behavior of a real system cannot be developed until one can see simpler models

---

[5]I have tried to carry on this tradition by contributing "Brain Twisters" to the *Daily Telegraph* and to the BBC Radio 4 program "Puzzle Panel".

within it. But what are these simpler models? They are generally among the large variety of models that have been created in the past, often for recreational or fanciful reasons.

Perhaps the clearest example is graph theory, where Euler made a simple model of the reality that he was studying, then later workers found that model useful in other situations. Graphs were then recognized as present in many early problems: river crossing in c. 800, knight's tours in c. 900, etc.

Thus, *recreational mathematics helps as a major source of mathematical models, techniques, and methods,* which are the raw material for mathematical research, in the same way that mathematics in general serves as a source of models for the physical world. This is the explanation of the utility of recreations in mathematics and perhaps the utility of mathematics in the real world.

Bibliography

[1] Alcuin (*c.* 735–804). *Propositiones Alcuini doctoris Caroli Magni Imperatoris ad acuendos juvenes.* 9th century. See companion volume for more.

[2] C. G. Bachet. *Problmes plaisans et délectables qui se font par les nombres.* (1612). The 5th ed., edited by A. Labosne in 1884, reprinted by Blanchard, 1959.

[3] W. W. Rouse Ball. *Mathematical Recreations and Essays.* Macmillan, London, 1892. 13th ed., Dover, NY, 1987.

[4] E. Baudot. "La Télégraphie multiple." *Annales télégraphiques (Ser. 3),* 22 (1870) 28–71, 152–177.

[5] E. Baudot. "Télégraphiques multiple imprimeur." *Annales télégraphiques (Ser. 3),* 6 (1879) 354–389.

[6] R. Borndörfer, M. Grötschel and A. Löbel. "Alcuin's transportation problems and integer programming," in *Karl der Grosse und sein Nachwirken. 1200 Jahre Kultur und Wissenschaft in Europa,* edited by Paul Leo Butzer, M. Kerner, and Walter Oberschelp, Brepols, 1997, 379–409.

[7] F. Calandri. *Aritmetica.* c. 1500. Edited by Gino Arrighi, Edizioni Cassa di Risparmio di Firenze, Florence, 1969.

[8] J. Cardan. (1501–1576). *Practica Arithmetice, & Mensurandi Singularis.* Bernardini Calusci, 1539 (often reprinted, e.g., in 1967).

[9] H. E. Julyan Cartwright and D. L. González. "Möbius strips before Möbius: Topological hints in ancient representations." *Mathematical Intelligencer,* 38(2) (2016) 69–76 (and front cover).

[10] *The Columbia Algorism*. Italian MS of c. 1370 at Columbia University. See companion volume for more.

[11] P. Diaconis and R. Graham. *Magical Mathematics*. 2012, Princeton University Press, 25–29, 42–60.

[12] P. Leonardo, called Fibonacci. (*c.* 1170–1240) *Liber Abbaci*, 1202. Translated by Laurence E. Sigler as *Fibonacci's Liber Abaci: A Translation into Modern English of Leonardo Pisano's Book of Calculation*. Springer, New York, 2002.

[13] C. Flye Sainte-Marie. "Question no. 48." *L'intermédiare des mathématiciens*, 1 (1894). (Proposal by A. de Rivière, 19–20; solution by Flye Sainte-Marie, 107–110.)

[14] A. F. Frey, Jr. and D. Singmaster. *Handbook of Cubik Math.*, Enslow Publishers, 1982 (also Lutterworth Press 2010).

[15] M. Gardner. "Penrose Tilings." *Penrose Tiles to Trapdoor Ciphers . . . and the Return of Dr. Matrix*, W. H. Freeman, 1997, 1–18. (Revised edition, Mathematical Association of America. 2005.)

[16] M. Gardner, "Penrose tiling II." in *Penrose Tiles to Trapdoor Ciphers . . . and the Return of Dr. Matrix*, W.H. Freeman, 1997, 19–30. (Revised edition, Mathematical Association of America. 2005.)

[17] M. Guggenberger. "The Gallo-Roman dodecahedron." *Mathematical Intelligencer*, 35, 4 (Winter 2013) 56–60.

[18] Gurudev. "World's oldest combinatoric formula." 2007. (Online at hitxp.)

[19] W. R. Hamilton. "Account of the Icosian Calculus." *Proc. Roy. Irish Acad.*, VI (1858) 415–416. Hamilton's collected papers (edited by H. Halberstam and R.E. Ingram) notes that "[t]his is the substance of a letter written on 27 October 1856 to the Rev. Charles Graves, D.D."

[20] F. G. Heath. "Pioneers of binary coding." *Journal of the Institution of Electrical Engineers* 7 (September 1961) 539–541.

[21] F. G. Heath. "Origins of the binary code." *Scientific American*, 227, 5 (May 1972) 76–83, 124.

[22] D. Huylebrouck. "Tourist anecdotes." *Mathematical Intelligencer*, 34, 4 (Winter 2012) 53–55.

[23] F. Josephus. ca. 75. *De bello Judaico* [The Jewish War]. Book III, chap. 8, sect. 7. Translation by H. St. J. Thackeray. Vol. 2, Heinemann, 1927, 685–687.

[24] D. A. Kerr. "Letters." *Scientific American*, 205, 1 (July 1961) 14.

[25] L. L. Larison. "The Möbius band in Roman mosaics." *American Scientist*, 61, 5 (September–October 1973) 544–547.

[26] Y. Li and S. Du. [*Chinese Mathematics: A Concise History*] (in Chinese). Hong Kong: Commercial Press, 1965. (English translation by John Crossley and Anthony W.-C. Lun. Clarendon Press, 1987.)

[27] O. Neugebauer. *Mathematische Keilschrift-Texte* (*"MKT"*). Springer-Verlag, 1935. (Reprinted by Springer, 1973.)

[28] K. H. Rosen. *Elementary Number Theory and Its Applications.* Addison-Wesley, 1984. (Also 6th ed., Pearson, 2011.)

[29] P. Singh. "The so-called Fibonacci numbers in ancient and medieval India." *Historia Mathematica*, 12 (1985) 229–244.

[30] D. Singmaster. *Notes on the "Magic Cube".* Privately published, 1979. (Also 5th ed., *Notes on Rubik's Magic Cube*, Enslow Publishers, 1981 and Penguin Books, 1981.)

[31] D. Singmaster. "The utility of recreational mathematics." in *The Lighter Side of Mathematics: Proceedings of the Eugène Strens Memorial Conference on Recreational Mathematics and its History*, edited by Richard K. Guy and Robert E. Woodrow, Mathematical Association of America, 1996, 340–345.

[32] D. Singmaster. "The utility of recreational mathematics." in *Teaching Mathematical Concepts Using Puzzles and Games*, Sheffield Hallam University Press, 2000.

[33] J. Slocum, D. Singmaster, W.-H. Huang, D. Gebhardt and G. Hellings. *The Cube: The Ultimate Guide to the World's Best-selling Puzzle — Secrets, Stories, Solutions.* Black Dog & Leventhal Publishers, 2009.

[34] S. K. Stein, "The mathematician as an explorer." *Scientific American*, 204 (5) (May 1961): 149–158, 206. (With revisions: "Memory Wheels." in *Mathematics, The Man-Made Universe.* (3rd ed.) W. H. Freeman, 1976, 141–155.

[35] E. Wigner, "The unreasonable effectiveness of mathematics in the natural sciences." *Communications in Pure and Applied Mathematics*, 13, 1 (February 1960), 1–14.

[36] C. E. Yü, (= C. E. Yu). *Ingenious Ring Puzzle Book.* (in Chinese) (Shanghai Culture Publishing Co., Shanghai, 1958.) Translated by Yenna Wu. (also *Ingenious Rings.* China Children's Publishing House, Beijing, 1999. edited into simplified Chinese by Lian Huan Jiu, with some commentary by Wei Zhang.)

Chapter 2

On Round Pegs in Square Holes and Vice Versa

Some problems just call out for attention. Which fits better, a round peg in a square hole or a square peg in a round hole? This can easily be solved once one arrives at the following mathematical formulation of the problem. Which is larger: the ratio of the area of a circle to the area of the circumscribed square or the ratio of the area of a square to the area of the circumscribed circle? One easily finds that the first ratio is $\pi/4 \, (= 0.785\ldots)$ and that the second is $2/\pi \, (= 0.636\ldots)$. Since the first is larger, we may conclude that a round peg fits better in a square hole than a square peg fits in a round hole.

The first person to pose this seems to be Bagley, in 1947 [5]: "Which is the worst misfit, a square peg in a round hole or a round peg in a square hole?" He shows the round peg fits better, but he only considers two dimensions. The above question leads to new problems. We naturally generalize the question to n dimensions. The remainder of this chapter will be devoted to higher dimensions.[1]

Generalizations of this sort are the hallmark of mathematical research. Like in all of science, over time investigations tend away from the specific to the general. The following result summarizes our results.

---

[1]This was my first published paper: *Mathematics Magazine*, 37(5) (November–December 1964), 335–337. Martin Gardner cited it in his *Scientific American* column [2].

Theorem 2.1. *The n-ball fits better in the n-cube than the n-cube fits in the n-ball if and only if $n \leq 8$.*

First, we take the following formula for the n-volume of the n-ball of radius r [1]

$$V = \frac{\pi^{\frac{n}{2}} r^n}{\Gamma\left(\frac{n}{2} + 1\right)},$$

where $\Gamma(x)$ is the well-known gamma function. The gamma function, which generalizes the factorial function, is discussed in this chapter's appendix. (It is noteworthy that the volume of the unit n-ball approaches zero with increasing n.)

Since we are interested only in ratios, we may, without loss of generality, assume that we have the unit n-ball in both ratios. Then the edge of the circumscribed cube is 2. Since the diagonal of an n-cube is $\sqrt{n}$ times its edge, we see that the edge of the n-cube which is inscribed in the unit n-ball is $\frac{2}{\sqrt{n}}$. Letting $V(n)$, $V_c(n)$, and $V_i(n)$ represent the n-volumes of the unit n-ball, its circumscribed n-cube, and its inscribed n-cube, respectively, we have

$$V(n) = \frac{\pi^{n/2}}{\Gamma(n/2 + 1)}, \quad V_c(n) = 2^n, \quad V_i(n) = \frac{2^n}{n^{n/2}}.$$

Now, the ratios under consideration are as follows:

$$R_1(n) = \frac{V(n)}{V_c(n)} = \frac{\pi^{n/2}}{\Gamma((n+2)/2)2^n},$$

$$R_2(n) = \frac{V_i(n)}{V(n)} = \frac{2^n \Gamma((n+2)/2)}{n^{n/2} \pi^{n/2}}.$$

The ratio $R_1(n)$ measures how well an n-ball fits in an n-cube and the ratio $R_2(n)$ measures how well an n-cube fits in an n-ball. Theorem 2.1 can now be stated as follows: $R_1(n) \geq R_2(n)$ if and only if $n \leq 8$. We shall prove somewhat more.

Theorem 2.2. $\frac{R_1(n)}{R_2(n)} \to 0$ *as* $n \to \infty$.

Proof. From the definitions, we have

$$\frac{R_1(n)}{R_2(n)} = \frac{\pi^n n^{n/2}}{2^{2n}\Gamma((n+2)/2)^2}.$$

We apply Stirling's approximation: $\Gamma(z) \sim z^{z-1/2}e^{-z}\sqrt{2\pi}$, thus obtaining

$$\Gamma((n+2)/2)^2 \sim \left(\frac{n+2}{2}\right)^{n+1} e^{-n-2}2\pi.$$

Hence, we have

$$\frac{R_1(n)}{R_2(n)} \sim \frac{\pi^n n^{n/2}}{2^{2n}\left(\frac{n+2}{2}\right)^{n+1}e^{-n-2}2\pi} = \frac{e^2\left(\frac{\pi e\sqrt{n}}{2(n+2)}\right)^n}{\pi(n+2)} < \frac{e^2\left(\frac{\pi e}{2\sqrt{n}}\right)^n}{\pi n}.$$

This last quantity is easily seen to approach zero as n increases. $\square$

Corollary 2.1. $R_1(n) < R_2(n)$ *for all large enough n.*

One may readily compute that the asymptotic approximation for $\frac{R_1(n)}{R_2(n)}$ has the value $1.06\ldots$ for $n=8$ and the value $0.84\ldots$ for $n=9$. Since Stirling's approximation has a relative error less than $\frac{1}{12z}$, we can say that the relative error in the asymptotic expression for $\frac{R_1(n)}{R_2(n)}$ is less than $2\cdot\frac{1}{12\cdot5}$ or 3.3% for $n\geq8$.

Hence we can be confident in stating that $R_1(n) < R_2(n)$ holds if $n\geq9$, since the asymptotic approximation decreases with n, for $n\geq5$. That the asymptotic approximation decreases with n is clear when $n\geq14$ since $\frac{\sqrt{n}}{n+2}$ decreases with n for $n\geq2$ and $\frac{\pi e\sqrt{n}}{2(n+2)} < 1$ for $n\geq14$. Further, one can compute that the asymptotic approximation decreases in the range $5\leq n\leq14$. Hence Theorem 1 has been half proven.

In order to check the theorem for small values of n, a program was written to compute $V(n), R_1(n)$ and $R_2(n)$ for $1\leq n\leq100$. The results, which are partially reproduced below, show that $R_1(n)\geq R_2(n)$ holds if $n\leq8$, with equality only for $n=1$. The numerical results for small n, together with the asymptotic results for large n, show that $R_1(n)\geq R_2(n)$ if and only if $n\leq8$, as originally claimed.

n	$V(n)$	$R_1(n)$	$R_2(n)$
1	2.0	1.0	1.0
2	3.14159	0.78540	0.63662
3	4.18879	0.52360	0.36755
4	4.93479	0.30842	0.20264
5	5.26378	0.16449	0.10875
6	5.16770	0.080745	0.057336
7	4.72475	0.036912	0.029853
8	4.05870	0.015854	0.015399
9	3.29850	0.0064423	0.0078861
10	2.55015	0.0024904	0.0040154
11	1.88410	0.00091997	0.0020350
12	1.33526	0.00032599	0.0010273
13	0.91062	0.00011116	0.00051691
14	0.59926	0.000036576	0.00025936
15	0.38144	0.0000011641	0.00012982
16	0.23533	0.0000035908	0.000064840
17	0.14098	0.0000010756	0.000032325
18	0.082145	0.00000031336	0.000016088
19	0.046621	0.000000088923	0.0000079952
20	0.025807	0.000000024611	0.0000039680
30	0.000021915	2.0410×10^{-14}	3.4146×10^{-9}
40	3.6047×10^{-9}	43.2784×10^{-21}	2.7741×10^{-12}
50	1.7302×10^{-12}	41.5367×10^{-28}	2.1835×10^{-15}
60	3.0962×10^{-18}	2.6856×10^{-36}	1.6844×10^{-18}

One can also show that $R_2(n)$ and $\frac{V(n)}{R_2(n)}$ each approach zero with $V(n) > R_2(n)$ if and only if $n \leq 61$.

2.1 Extremal Spheres

About 1964, John Kelly,[2] observed that there must be a dimension between 8 and 9 where $R_1(n) = R_2(n)$. Indeed, the expressions are all continuous and differentiable functions of n, so this follows from the Intermediate Value Theorem. Herman P. Robinson kindly computed it, in 1979, and obtained $n = 8.137941046091372\ldots$.

---

[2]Kelly was one of my teachers at Berkeley.

In 1979, the following problem was posed: Determine the unit sphere of maximum volume (with respect to dimension). The published solution [4] only considered integral dimensions, finding the maximum as $V(5) = \frac{8}{15}\pi^2$. An editorial note says R. P. Boas recalled seeing a graph of $V(n)$ when he was a student, but he could not recall where. Herman P. Robinson kindly calculated the value of n for me, getting

$$n = 5.256946404860576\ldots, \quad V(n) = 5.277768021113400\ldots.$$

A little more consideration of n-dimensional geometry gives results for the surface area of the unit sphere in n dimensions as $A(n) = nV(n) = 2\pi V(n-2)$, so the maxima of A and of V occur at dimensions just two apart. Hence the maximal surface area of an n-sphere occurs at

$$n = 7.256946404860576\ldots,$$

$$A(n) = 2\pi \times 5.27\ldots = 33.161194489301048\ldots.$$

2.2 Popular Conceptions

Since this chapter first appeared in 1964, quotations about the topic have been collected.[3]

- "You gentlemen", said he, "are like square handles which you would thrust into the round sockets of your generation. Consequently, there is not one of you which fits".

Wang Tao-K'un. *How to get on.* Late 16C. Translated excerpt in: *Gems of Chinese Literature*, 2nd ed., ed. by Herbert A. Giles, Kelly & Walsh, 1923; reprinted by Dover, 1965, p. 227.

- If you choose to represent the various parts in life by holes upon a table, of different shapes — some circular, some triangular, some square, some oblong — and the persons acting these parts by bits of wood of similar shapes, we shall generally find that the triangular person has got into the square hole, the oblong into the triangular, and a square person

---

[3]A version of this section appeared as my "Letter: Quotations on round pegs in square holes, etc.", *Theta* 5(1) (1991) 36. I have expanded it since. I would be very grateful for further quotations.

has squeezed himself into the round hole. The officer and the office, the doer and the thing done, seldom fit so exactly that we can say they were almost made for each other.

Sydney Smith (1771–1845). *Sketches of Moral Philosophy.* Lecture IX, 1824. Requoted in Hesketh Pearson, *The Smith of Smiths*, The Folio Society, 1934, p. 92, and in many other places.

- Sir Robert Peel was a smooth round peg, in a sharp-cornered square hole, and Lord Lyndhurst is a rectangular square-cut peg, in a smooth round hole.

Fonblanque, *England under Seven Administrations*, 1837, vol. III, p. 342; quoted in the *The Oxford English Dictionary* (OED) under "peg".

- Was there ever a more glaring case of square peg in round hole and round peg in square?

From the *Westminster Gazette* of 24 December 1901, found in the OED.

- . . . Instead of finding fit offices for fit men, we are perpetually discovering, on the stage of society, actors out of character! Our most popular writer has happily described this error. "A leading philosopher, the Democritus of our day, once compared human life to a table pierced with a number of holes, each of which has a pin made exactly to fit, but which pins being stuck in hastily, and without selection, chance leads inevitably to the most awkward mistakes. For how often do we see", the orator pathetically concluded, "how often, I say, do we see the round man stuck into the three-cornered hole!"

Isaac D'Israeli. *Curiosities of Literature*, New ed. in one volume, Routledge, 1867. "Anecdotes of the Fairfax family". p. 348.

- The old proverb about the fit of a square peg in a round hole must be revised to include the lack of fit of our five-sided arithmetic in our foursquare minds.

E. M. Tingley. "Calculate by eights, not by tens". *School Science and Mathematics*, April 1934, 395–401. Quoted by Underwood Dudley, *Mathematical Cranks*, Math. Assoc. of America, 1992, 28.

- Indeed, it is too much to expect of Providence that the square peg and the round hole will never be brought together; . . .

A. P. Herbert. *Uncommon Law*, Methuen, 1935, (new ed., 1969); Eyre Methuen, London, 1977. Case 66: Pale, M. R., v. Pale, H. J., and Hume (The King's Proctor showing cause). "Not a Crime". 425–458.

- One on the outside who criticizes the placement of square pegs in round holes should be sure that there are not more round holes and square pegs than there are square holes and round pegs. Even if this is not the case the critic should be certain that round holes are not a more serious problem than square ones, and he should withhold his criticism unless he is quite sure that it is better to leave round holes unfilled than it is to fill them partially with square pegs.

American Journal of Public Health. Not further identified. Quoted in: Denys Parsons; *True to Type*; Macdonald, London, 1955, p. 23. Republished in his *The Best of Shrdlu*; Pan, 1981, p. 91.

- Whether each circle is round,
  Is a question both deep and profound.
  In a paper of Erdös,
  Written in Kurdish,
  A counterexample is found.

Attributed to Leo Moser (1921–1970), but a reference has not been tracked down.

- He who is born square will not die round.

Neapolitan proverb. Given in Bruno Munari, *Discovery of the Square*, Alec Tiranti, 1965, 59 and in his *Square Circle Triangle*, Princeton Architectural Press, 2015, 65.

- Square circles or round squares have haunted many diverse philosophical writers as the archetype of the impossible and the absurd: they were assigned a place near — or rather below — golden mountains, unicorns and mermaids.

Karl Menger. "Square circles (The taxicab geometry)". in *Selected Papers in Logic and Foundations, Didactics, Economics*, Reidel, 1979. Quoted by Louise Golland, "Karl Menger and taxicab geometry", *Amer. Math. Monthly* 63, 1990, 326–327.

- The phrase "square peg in a round hole" comes from the fact that only square pegs were used in round holes, a round peg was no use, they fell out. All medieval and later wooden building were fixed so. All "joined furniture" up till the use of

> animal glue by cabinet makers, (*c.* 1700) had square pegs in round holes. — Yours, etc. (Professor) Stephen Baty, Institute of Antiquarian Crafts

Letter in the *Irish Times*, 28 July 1995, 17.

- > Mrs. Johnston, of Bradford, says her son doesn't know what profession to take up. Can the Brains Trust suggest what signs to look for if a parent is to try to avoid putting a square peg in a round hole?

Question put to "Any Questions?", generally known as the "Brains Trust", BBC radio, c. 1942, in Howard Thomas, ed. *The Brains Trust Book*, Hutchinson, c. 1942, 90.

- > "Or were the police fitting a very square peg into a very round hole?"

William Beadle. *The Killing of Leon Beron.* Wat Tyler Books, 1995, p. 68.

- > "Diruit, aedificat, mutat quadrata rotunda" [He pulls down, he builds up, he changes square things into round].

Horace. Quoted in C. O. Sylvester Mawson, *Le Mot Juste: The Penguin Dictionary of Foreign Terms and Phrases*, 1934 (and 1987, 1988, 1990), 99.

In addition to these quotes the imagery is used in cartoons and other places. It is an obvious idea for a cartoon. Several have been collected but we present only two here, Figures 2.1 and 2.2.

The distinguished Dutch puzzle inventor, Oskar van Deventer, makes a board with a circular and a rectangular hole and a piece of a circular cylinder and a rectangular block. The piece of circular cylinder has diameter too large to pass through the circular hole and the block is too long to fit through the rectangular hole. One has to turn the pieces and pass the block through the circular hole and the cylinder through the rectangular hole. Can a three hole and three piece version can be designed?

2.3   Educational Value

The problem appeared on a SMILE (Secondary Mathematics Individualized Learning Experiment) card. This experiment started in the

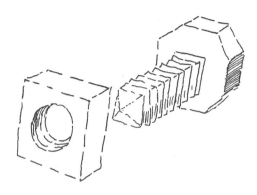

Figure 2.1. Ed Powers, illus. Arthur Bloch, *The Complete Murphy's Law.*, Price/Stern/Sloan, 1990, 109.

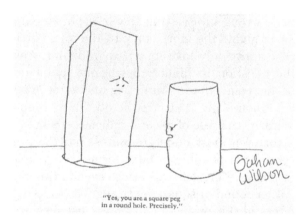

Figure 2.2. Gahan Wilson. "*... and then we'll get him!.*" Richard Marek, 1978, 116.

1970s and produced problems to encourage secondary students to learn how mathematics is done. A letter to *ILEA Contact* (a weekly newsletter of the Inner London Education Authority) criticized the design and production of SMILE cards and cited this "seemingly daft question." My response appeared on 12 September 1980, and the points expressed are still valid.

Sir — The letter in the 4 July issue complaining about a SMILE card which asked whether a round peg fits better in a square hole than a square peg fits in a round hole brought a smile (!) to my face. My first published paper was "On round pegs in square holes and square pegs in round holes" (*Mathematics Magazine* 37 (1964), pp. 335–337) and I imagine that the SMILE question traces back to my paper.

The problem arose because I could never remember which way round the proverb went. I have just looked in Brewer's *Dictionary of Phrase and Fable* (1970 edition) and I find the main entry is under Peg, referring to a square peg in a round hole. However, there is also an entry under Round, referring to a round peg in a square hole as "The same as a square peg in a round hole."

In my paper, I gave a mathematical formulation of the question which readily yields that the respective "ratios of fit" are $\pi/4$ and $2/\pi$ so that a round peg fits better in a square hole than vice versa. Hence Brewer is wrong in asserting that the phrases are the same, but he is right in giving the main entry to the worse fitting case.

However, the above only takes one paragraph of my paper. I then considered the question in higher dimensions and I obtained the rather unexpected result that a round peg fits better in a square hole only up to dimension 8 and that it is the other way round for higher dimensions. Clearly, the case of general dimension is not accessible to school pupils (or even most undergraduates) but the generalization to three dimensions is possible in later forms [i.e., years]. In earlier forms, there are interesting generalizations even in two dimensions — e.g., replace either round or square by triangular. Indeed, I have realized that I never did this problem and I have just shown that a circle fits better in a regular n-gon than vice versa for all n greater than or equal to 3.

The problem with two polygons gets very messy. Readers might like to find the largest equilateral triangle which can fit into a square or the largest square which can fit into an equilateral triangle. Masochistic readers may like to consider arbitrary triangles or quadrilaterals or to consider pentagons, hexagons, etc. In three dimensions, there are only five regular solids, so there are only five problems with one round (i.e., spherical) shape and 20 problems with regular solids. I have not yet computed the ratios in the former cases, though they are fairly straightforward. In the latter cases, it is not yet known what the largest regular solid inside another one is in several of the problems.

I hope that this letter will indicate some of the ways in which the problem can stimulate and interest some people. I must admit that most of this is beyond second year, but the realization of a mathematical concept in a situation which is not immediately mathematical, the formulation of the concept, the solution of the formulated problem and the consideration of generalizations in the plane are all more or less at this level. Further, the more complicated generalizations may be an incentive to learn trigonometry and solid geometry.

David Singmaster

2.4   Appendix

The Gamma Function

The Gamma Function does not come up in recreational mathematics very often, but it does occur throughout analytic number theory and probability theory, so readers might appreciate a note on it. Basically, the Gamma Function is the generalization of the factorial function to non-integral arguments. It is well known that

$$\int_0^\infty e^{-t}t^n dt = n!.$$

In considering this in 1729, Euler found it was convenient to shift the index and define

$$\Gamma(x) = \int_0^\infty e^{-t}t^{x-1}dt.$$

Since this is an improper integral, one has to check that it converges and it does for positive x. Integrating by parts gives us

$$\Gamma(x+1) = x\Gamma(x) \tag{2.1}$$

and this can be used to define the function for negative real numbers. We also have

$$\Gamma(1) = 1 \tag{2.2}$$

and hence $\Gamma(n) = (n-1)!$ for positive integers n. However, there are many functions satisfying Eqs. (2.1) and (2.2). In 1922, Harald Bohr (brother of Niels Bohr) and Johannes Mollerup proved that Euler's definition was the only function satisfying Eqs. (2.1) and (2.2) which was appropriately smooth, i.e., had a convex logarithm. It is remarkable that $\Gamma(\frac{1}{2}) = \sqrt{\pi}$. Emil Artin, wrote a classic small monograph on it, recommended to anyone who has studied any analysis [3].

Bibliography

[1]   D.M.Y. Sommerville, *An Introduction to the Geometry of N Dimensions*. Dover, 1958, 136.

[2]   M. Gardner, "Spheres and Hyperspheres." *Mathematical Circus*. Knopf, 1979, 39. This reprints the original May 1968 column.

[3]   E. Artin, *Einführung in die Theorie der Gammafunktion*. Teubner, Leipzig, 1931. Translated by Michael Butler as *The Gamma Function*. Holt, Rinehart and Winston, 1964.

[4]   D. Singmaster, "Problem 80–14 An Extremal Sphere." *SIAM Review*, 23(3) (July 1981) 394–395.

[5]   W. A. Bagley, *Paradox Pie*. Vawser & Wiles, c. 1947. No. 17: "Misfits," 18.

Chapter 3

Hunting for Bears

> A hunter travels 10 miles south, then 10 miles east and then 10 miles north. He finds himself at his starting point. He sees and shoots a bear. What color was the bear?

Most readers will have heard this famous bear hunting problem. The standard answer is "white" because he can travel in the stated way if and only if he starts at the North Pole.[1]

Geographic problems of this sort go back to the early 18th century and probably earlier. The earliest versions ask where can one have a house with all windows facing south or where the wind always blows from the south? In 1925, Ackermann [1] asked: a man at the North Pole goes 20 miles south and then 30 miles west. How far, and in what direction, is he from the Pole? In 1933, Hubert Phillips [7] asked: If you start at the North Pole, go 40 miles South, then 30 miles West, how far are you from the Pole? Answer: "Forty miles". (*Not* thirty, as one is tempted to suggest.)[2] The earliest bear problem seen is from 1943 [4]. Surprisingly, a four-sided version occurs earlier, in Perelman (see below).

There are other locations from which you can travel 10 miles south, 10 miles east and then 10 miles north and find yourself at your starting point. If you start near the South Pole and your eastward journey circles the Pole one (or more) times, you can return on the

---

[1] Adapted from "Square Hunting" in "Symmetry Saves the Solution". in Alfred S. Posamentier & Wolfgang Schulz, eds., *The Art of Problem Solving*, Corwin Press, 1996, 276–278.

[2] He probably meant to use fifty rather than thirty.

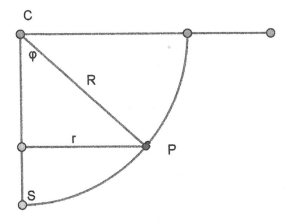

Figure 3.1.

same meridian for your northward return that you went along on your southward stage. However, there are no bears near the South Pole, so this does not produce answers for the color question.

Let P be one of these locations. Let Φ be its co-latitude, i.e., the angle from P to the center of the earth to the South Pole. Observe that the circumference of the circle at this co-latitude Φ, is $2\pi r$, where $r = R \sin \Phi$ and R is the radius of the earth. See Figure 3.1.

The 10 mile eastward journey at this co-latitude is a multiple of the circumference at this co-latitude if $10 = 2k\pi R \sin \Phi$ for some integer k. Setting Φ_k as this angle, we have $\Phi_k = \sin^{-1} \frac{10}{2\pi k R}$. Taking $k = 1$, we get $\sin \Phi_1 = \frac{1}{800\pi}$, which gives $\Phi_1 = 0.0228°$, so we are pretty close to the South Pole.

3.1 The Square Path Version

An explorer travels on the surface of the earth, assumed to be a perfect sphere, in the manner to be described. First, he travels 100 miles due north. Then he travels 100 miles due east. Next he travels 100 miles due south. Finally, he travels 100 miles due west, ending at the point where he started. Determine all the possible points from which he could have started (see Figure 3.2) [3]. The first solver gave only the obvious solution: 50 miles south of the equator. The later solver found the general solution but could not find the answers in

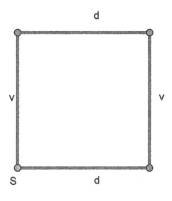

Figure 3.2.

closed form. In 1983–1984, the problem and its solution appeared in *The Journal of Recreational Mathematics* [2]. Barwell found a better solution but did not quite carry it to its conclusion. This is probably based on a problem of Perelman, ca. 1934 [5] which asked where one would be if one started from Leningrad and made such a journey in a dirigible, with $500\,\mathrm{km}$ sections — a later version, in 1985, uses a helicopter [6]. The following closed form solution, using a standard symmetry, is not found in the literature.

We generalize somewhat, as shown in Figure 3.2. We start at S and go v to the north, then d to the east, v to the south and d to the west, and return to our starting point. Let the polar angle (= co-latitude) of the eastward journey be Φ_1 and the polar angle (co-latitude) of the westward journey be Φ_2. We have

$$\Phi_2 - \Phi_1 = \frac{v}{R}. \tag{3.1}$$

If we are near the North Pole, the westward journey can encircle the pole and then we want

$$\frac{d}{R \sin \Phi_1} \equiv \frac{d}{R \sin \Phi_2} \ (\mathrm{mod}\ 2\pi) \tag{3.2}$$

or

$$\frac{d}{R \sin \Phi_1} = \frac{d}{R \sin \Phi_2} + 2\pi k \tag{3.3}$$

for some integer k, which is the number of times the explorer circles the earth going east. The solution of Eq. (3.1) and Eq. (3.2) or

Eq. (3.3) is not obvious and frustrated the 1960 solver. Assuming we are not at the special cases, $\Phi_1 = 0$ or $\Phi_2 = \pi$, we can symmetrize the problem by setting

$$\Phi = \frac{\Phi_1 + \Phi_2}{2}, \quad \beta = \frac{\Phi_1 - \Phi_2}{2},$$

so

$$\Phi_1 = \Phi - \beta, \quad \Phi_2 = \Phi + \beta.$$

Barwell [2] does the same. Setting $\alpha = \frac{d}{2R}$ and using trigonometric identities, Eq. (3.3) becomes

$$\frac{\pi k}{\alpha} = \frac{1}{\sin(\Phi - \beta)} - \frac{1}{\sin(\Phi + \beta)} = \frac{\sin(\Phi + \beta) - \sin(\Phi - \beta)}{\sin(\Phi - \beta)\sin(\Phi + \beta)}$$

$$= \frac{2\cos\Phi\sin\beta}{\frac{1}{2}\cos 2\beta - \frac{1}{2}\cos 2\Phi} = \frac{2\cos\Phi\sin\beta}{cos^2\beta - \cos^2\Phi}.$$

Assuming again that we are not in the special case $k = 0$, we can continue. Cross-multiplying the above equation gives a quadratic equation whose solution is

$$\cos\Phi = \frac{-\alpha\sin\beta \pm \sqrt{\alpha^2\sin^2\beta + \pi^2 k^2\cos^2\beta}}{\pi k}. \tag{3.4}$$

We have assumed $k \neq 0$. The case $k = 0$ leads to $\Phi_1 = \Phi_2$ or $\Phi_1 + \Phi_2 = \pi$ with $\Phi = \frac{\pi}{2}$. In our derivation of Eq. (3.4) we have two square roots involved. Taking the negative value of $\cos\Phi$ simply moves things to the other side of the globe. But taking the negative square root (or square route!) in Eq. (3.4) leads to solutions which cross over the Pole and travel in the directions N, W, N, W. We can change this to going N, E, N, W, but this changes the sign in Eq. (3.3), leading to a reversal of sin and cos in Eq. (3.4). Using the positive square root in Eq. (3.4) then leads to solutions not crossing the pole but going N, W, S, W.

Barwell [2] got to the quadratic but thought the solution too messy to consider and went on to consider approximate solutions. This solution leads to a number of further questions and variations of the problem. Some of these are discussed below. None of this would have been possible without the symmetrization step. This process goes back to the Babylonians, who used it to solve such problems as $x + y = 10, xy = 24$ by setting $x = 5 + \beta, y = 5 - \beta,$

so $xy = 25 - \beta^2 = 24$ and $\beta^2 = 1$. This symmetrization has the effect of eliminating the first-order term in a quadratic and is equivalent to completing the square. The square bear problem is a trigonometric problem and hence more complicated, but the symmetrization is even more essential. An example of a new problem inspired by the above is as follows:

> Hiawatha, the mighty hunter, has traveled far in search of game. One morning he gets up, has breakfast, heads north and travels 10 miles forward in a straight line. Seeing nothing, he stops for lunch. After lunch, he heads north and travels 10 miles forward in a straight line and finds himself where he started in the morning. Where on earth is he? See [8, 9].[3]

Bibliography

[1] A. S. E. Ackermann. *Scientific Paradoxes and Problems and Their Solutions.* The Old Westminster Press, 1925, 116.

[2] Y. Kakinuma, B. Barwell and C. H. Collins. "Problem 1212: Variation of the Polar Bear Problem." *Journal of Recreational Mathematics.*, 16, 3 (1984–1985) 226–228.

[3] M. Klamkin, D. A. Breault and B. L. Schwartz. "Problem 369." *Mathematics Magazine*, 32 (1958/1959) 220. (Also "The Explorer." 32, 4 (November–December 1959/1960) 110; and "The Explorer Revisited." 33, 4 (March–April 1959/1960) 226–228.)

[4] E. J. Moulton. "A Speed Test Question. A Problem in Geography." *American Mathematical Monthly*, 51 (1944) 216 & 220.

[5] Y. I. Perelman. *Figures for Fun.* Frederick Ungar, 1965. (Also Dover 2015.) (Original in Russian, *Zhivaia Matematika*, 1934.)

[6] Y. I. Perelman. *Mathematics Can Be Fun.* MIR, 1957, problem 6, 14–15 & 19–20. (Original in Russian, *Zhivaia Matematika*, 1970, and *Zanimatelnaia Algebra*, 1976.)

[7] H. Phillips. *The Playtime Omnibus.* Faber & Faber, 1933. section XVI, "Problem 11: Polar conundrum," 51 & 234.

[8] D. Singmaster. "Symmetry Saves the Solution." in Alfred S. Posamentier & Wolfgang Schulz, eds.; *The Art of Problem Solving.* Corwin Press, 1996, 273–286.

---

[3]I have given the Hiawatha problem to several colleagues who have used it; here are some appearances of it.

[9] K. Devlin. "Explorer's problem." *The Guardian*, (18 June 1987) 16 & (2 July 1987) 16. (Also David Singmaster. "A walk on the wild side." *Games*, 15, 2 (August 1991) 57 & 40, and other places.) (And also Dick Hess. "Prob. 46: Where on earth?" *All-Star Mathlete Puzzles*, Sterling, 2009, 23 & 66.)

Chapter 4

Sum = Product Sequences

Whether we stopped to think about it or not, we are all familiar with $2 + 2 = 2 \cdot 2$ and $1 + 2 + 3 = 1 \cdot 2 \cdot 3$ as examples where a sum of positive integers is equal to their product. It is a curious fact and we wonder if such examples are scarce. Andrew Palfreyman [1] discusses such sequences with equal sums and products.[1] As will be seen, there are a number of unsolved questions that arise and these led to some correspondence with Mike Bennett and Andrew Dunn and an article in the *Fibonacci Quarterly* (1995).

To be more formal, if $a_1, a_2, \ldots, a_n$ is a sequence of n positive integers in non-decreasing order such that $\sum a_i = \prod a_i$, we call it a Sum = Product sequence of size n. First, we summarize the known results. For notational convenience, we denote repeated values by an exponent — e.g., 2, 2, 2 will be denoted 2^3. A little calculation shows that the following table gives all Sum = Product sequences of size ≤ 11.

Proposition 4.1. *For every n, there is a Sum = Product sequence of size n.*

Proof. $1^{n-2}, 2, n$ works. $\square$

Proposition 4.2. *If $n \not\equiv 0 \,(\mathrm{mod}\ 6)$ and $n > 6$, then there are two or more Sum = Product sequences of size n.*

---

[1] This chapter appeared in *Symmetry Plus*, 53 (Spring 2014), 7–8. I used the idea in my *Weekend Telegraph* "Brain Twister" on 9 and 16 April 1988 [2]. I mentioned these in the Problem Session of the *Fifth International Conference of Fibonacci Numbers and their Applications* at St. Andrews. Scotland, in 1992.

n	sequence(s)
2	2^2
3	1, 2, 3
4	1^2, 2, 4
5	1^3, 2, 5; 1^3, 3^2; 1^2, 2^3
6	1^4, 2, 6
7	1^5, 2, 7; 1^5, 3, 4
8	1^6, 2, 8; 1^5, 2^2, 3
9	1^7, 2, 9; 1^7, 3, 5
10	1^8, 2, 10; 1^8, 4^2
11	1^9, 2, 11; 1^9, 3, 6; 1^8, 2^2, 4

Proof. The following three sequences give a Sum = Product sequence distinct from that of Proposition 4.1 in all the cases except those stated.

$$1^{n-2},\ 3,\ \frac{n+1}{2}\text{for } n \equiv 1\,(\text{mod } 2),\ n \geq 5.$$

$$1^{n-2},\ 4,\ \frac{n+2}{3}\text{for } n \equiv 1\,(\text{mod } 3),\ n \geq 10.$$

$$1^{n-3},\ 2^2,\ \frac{n+1}{3}\text{for } n \equiv 2\,(\text{mod } 3),\ n \geq 5.$$

$\square$

Proposition 4.3. *If $n \not\equiv 0$ or 24 (mod 30) and $n > 6$, then there are two or more Sum = Product sequences of size n.*

Proof. From Proposition 4.2, there are only the following residue classes to consider (mod 30): $0, 6, 12, 18, 24$. The following three sequences give a Sum = Product sequence distinct from that of Proposition 4.1 in all cases except those stated.

$$1^{n-2},\ 6,\ \frac{n+4}{5}\text{for } n \equiv 1\,(\text{mod } 5),\ n \geq 11.$$

$$1^{n-5},\ 2^4,\ \frac{n+3}{15}\text{for } n \equiv 12\,(\text{mod } 15),\ n \geq 12.$$

$$1^{n-3},\ 2,\ 3,\ \frac{n+2}{5}\text{for } n \equiv 3\,(\text{mod } 5),\ n \geq 8.$$

$\square$

The process of Propositions 4.2 and 4.3 can be extended. Consider a sequence: $1^a, b_1, b_2, \ldots, b_k, c$. Let $S_1 = b_1 + \cdots + b_k$ and $P_1 = b_1 \cdots b_k$. Note that $n = a + k + 1$. This gives us a Sum = Product sequence of size n if $a + S_1 + c = P_1 c$, which leads to $c = (a + S_1)/(P_1 - 1)$. This holds if and only if $a \equiv -S_1 \pmod{P_1 - 1}$ or $n \equiv k + 1 - S_1 \pmod{P_1 - 1}$. By letting $P_1 - 1$ vary over all divisors of a modulus m and taking all factorizations of $P_1 - 1$, we find solutions for various residue classes.

We wrote a program to carry this out for a given m and to determine the density of residue classes for which the above process did not generate a Sum = Product sequence distinct from those given by Proposition 4.1. Adding a new prime factor to m reduces the density, but we could find no modulus for which the density got down to 0. For $m = 210$, the density was 0.03333 and for $m = 2310$, the density was 0.02121. Increasing the power of a prime appearing in m generally does not affect the density, but 125 gives a lower density than 25.

Proposition 4.4. *If $n - 1$ is not prime, then there are two or more Sum = Product sequences of size n.*

Proof. If $n - 1 = ab$, with $1 < a \le b$, then $1^{n-2}, a + 1, b + 1$ is a Sum = Product sequence distinct from that in Proposition 4.1. $\square$

Corollary 4.1. *The density of integers n having only one Sum = Product sequence is zero.*

Let $N(n)$ be the number of different Sum = Product sequences of size n.

Corollary 4.2. $N(n) \ge \lfloor \{d(n-1) + 1\}/2 \rfloor$, *where $d(n-1)$ is the number of divisors of $n - 1$.*

Corollary 4.3. $\limsup_{n \to \infty} N(n) = \infty$.

Our program found that $N(n) = 1$ when $n = 2, 3, 4, 6, 24, 114, 174, 444$ and for no other values of $n < 1,440,000$. We also find that $N(n) = 2$ holds for 49 values of n up to 120,000, the largest being $n = 6174$ and 6324 and that $N(n) = 3$ holds for 78 values of n up to 120,000, the largest being $n = 7220$ and 11874.

Conjecture 4.1. $N(n) > 1$ *for* $n > 444$, $N(n) > 2$ *for* $n > 6324$, *and* $N(n) > 3$ *for* $n > 11874$.

Conjecture 4.2. $N(n) \to \infty$.

After the initial work on this topic, Erdös referred us to Schinzel, who provided several earlier references, going back to 1956, but these gave no new insights. Perhaps readers can extend the calculations and make some progress on the conjectures.

Bibliography

[1]　A. Palfreyman. "Products and Sums." *Symmetry Plus* 52 (Autumn 2013) 12–13.

[2]　B. R. Clarke, R. Gooch, A. Newing and D. Singmaster. *The Daily Telegraph Book of Brain-Twisters* No. 1. Pan Books, 1993. Prob. 2, 3 & 74 & 114.

Chapter 5

A Cubical Path Puzzle

5.1 The Original Puzzle

There are several versions of this puzzle on the market. Figure 5.1 shows an example.[1]

They are simple geometric analogues of "transformer" toys. Each has a chain of cubes, strung on an elastic cord. Each cube has a hole, either straight through or from a face to an adjacent face (effectively a right-angle bend). For a string of 27 cubes, an obvious task is to make a $3 \times 3 \times 3$ cube.

We first introduce a version of the puzzle made in Germany in the following manner. Some of the cubes have holes straight through them (S pieces) and some of them bend, having holes passing diagonally from a face to an adjacent face (B pieces), as shown in Figure 5.2.

In the particular puzzle manufactured, the sequence of Ends, Straights and Bends is as follows:

ESBSBSBSBBBBSBSBBBSBBSBBBSE.

(The End pieces are actually Straights but they could be Bends without changing their behavior.) Consequently, one can lay out the pieces in a plane in the pattern of Figure 5.3 where the dotted line is the path of the elastic cord.

---

[1]This was submitted to the *Journal of Recreational Mathematics*, in 1980 and 1985, but I never had a response.

Figure 5.1.   A cubical path puzzle. Photo courtesy of Eryk Vershen.

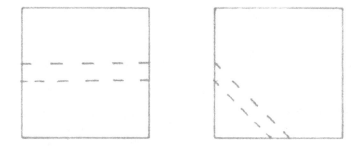

Figure 5.2.   Straight and Bend cubes.

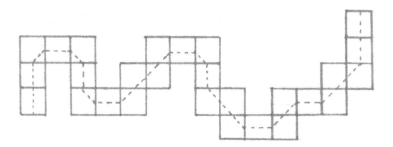

Figure 5.3.   A layout in the plane.

The puzzle arrives packed up into a $3 \times 3 \times 3$ cube. Only a little joggling is required for the cube to become unfolded. The puzzle is to put it back! (As an exercise: show there is only one way to make up a cube from this puzzle.)

Beyond this particular sequence/version many questions come to mind. What can one say about the number of straight and bend pieces in such puzzles? Could one have a 27-cube string with all bends? Many other problems are in the next section.

First, how many possible sequences are feasible?[2] Fortunately, the sequence of pieces had been recorded — otherwise there would have been $\binom{27}{11} = 13037895$ ways to arrange the pieces (actually $\binom{25}{9} = 2042975$ since the Ends were known to be S's). Some of these ways are clearly impossible — e.g., one cannot have two S's in a row. (This requirement reduces the number of possibilities to $\binom{17}{9} = 24310$ since the 9 S's can be placed in any set of 9 of the 17 spaces between or at the end of a row of B's.) Some sequences may appear possible but turn out impossible. On the other hand, some sequences will probably lead to many ways of making a cube. Answering such questions apparently involves exhaustive (and exhausting) searches.

Let's turn to a simpler question — how many S's can there be? In particular, could there be no S's? Let s be the number of S's and $b = 25 - s$ be the number of B's. In the following section, it is shown that $2 \leq s \leq 11$, i.e., $14 \leq b \leq 23$.

Maximum of s

We first observe that, since no 2 S's can be adjacent, $s - 1 \leq b$. Thus $24 = s + b - 1 \leq 2b$ and $b \geq 12$, $s \leq 13$.

If $s = 13$, then $b = 12$ and we must have the sequence:

$$\text{ESBSBSBSB}\ldots\text{SBSE.}$$

Examining the ways in which the initial part ESBSBSBSB can be placed in a cube, we see that the E and the B's must all be at corners and the succeeding B's must also be at corners. This gives 14 pieces to put at the 8 corners! So this cannot be done. Further we see that we cannot have more than 7 S's in a sequence BSBSB...BSB.

Now suppose $s = 12$, $b = 13$. Since each S must be separated from the next S by a B, we have only 2 B's extra and so we can have at most three successive B's. Thus we can have at most two double bends (i.e., ...SBBS...) or one triple bend (i.e., ...SBBBS...) or we may have the extra bend or bends placed next to the Ends. Consider the center cell of the $3 \times 3 \times 3$ cube. If this is occupied by a B then

---

[2]This came up when my 3 year old daughter pulled the elastic loose at one end.

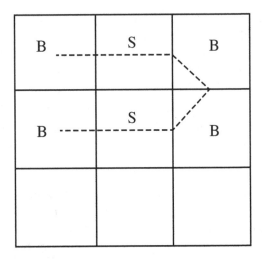

Figure 5.4.   The next piece must be a corner B piece above or below the upper left corner.

the two adjacent pieces in the sequence must be B's and so must the two pieces adjacent to those. This gives 5 or 4 B's in a row, except if we have EBBBS... where the first B is the center cell. But this puts three B's before the first S, leaving only 10 to fill in between the 12 S's. Hence the center cannot be a B.

Suppose now the center is an S. Again the two pieces adjacent to it must be B's, as must the two pieces adjacent to these. Consider the sequence SBB with S at the center. The last B is at an edge of the cube. Either the next piece is at a corner and is a B or the next piece is an S followed by 2 B's; see Figure 5.4 for the center layer. In either case, we can only make this work when the piece before the center S is an E. But then we will later have a sequence BSB...BSB with 9 or 10 S's which is impossible.

Finally we note that the center can never be an E. If we color the 27 cells of the cube by the parity of the sum of their coordinates, i.e., like a chessboard, we see that the 8 corners and 6 face centers are of one parity and the 12 edges and cube center are of the other parity. Our sequence of cells alternates between parities, so it must begin and end with cells of the first parity. (Incidentally, this shows that we cannot do the puzzle if all the 27 pieces are joined in a circuit.) Hence $s = 12$ is impossible.

Minimum of s

If we are at a corner, two B's can only bring us to a face center via one edge, and if we are at a face center, two B's can only bring us to a corner via an edge or to another face center via the cube center. Consider now the alternate sequence of 14 pieces at positions: $1, 3, 5, \ldots, 27$. These are the pieces in cells of the first parity. If $s = 0$ and we have all B's, then we must have a face center between each two corners in our sequence of 14. But this requires 7 face centers, so $s = 0$ is impossible. If $s = 1$, we must use the S to join two corners via an edge. The remaining corners can then be alternated with the 6 face centers. But each corner–face center–corner sequence uses two edges and so we need 13 edges, thus $s = 1$ is impossible. (As an exercise: Find solvable examples for $2 \leq s \leq 11$.)

5.2 Further Problems

How does one go about solving such problems systematically? Is there more than one solution? If so, how many solutions are there? The original version of this had one solution, but a set of five different versions was later marketed. Can you have versions which make several solid (or even plane) shapes? The problem of folding the path into a cube is equivalent to finding a "hamiltonian path" — that is, a path passing through each of the 27 "cubelets" of the cube exactly once — with the turn at each step specified in advance as either straight or a bend.

Scherphuis [3] lists all possible $3 \times 3 \times 3$ possible path puzzles that are "doable". This duplicates and corroborates the same list by Eryk Vershen [4]. Not counting rotations or mirror images of the hamiltonian paths, there are 11,487 puzzles, of which 3658 have unique solutions and one has 142 solutions. There are other web pages that come and go. Scherphuis shows that it is impossible to have a doable $3 \times 3 \times 3$ puzzle with only turns (see also Ruskey and Sawada [2]), and his webpage points to various pages with solutions for commercially available puzzles as well as for the Kibble Cube, a variation in which the cubes have grooves that allow for greater freedom.

There are marketed generalizations with the string forming a loop. This cannot be done for the $3 \times 3 \times 3$ form. For example, there is one with 36 cubes in a loop, with all pieces being bends, and making this into a $3 \times 3 \times 4$ takes a little effort. There are also examples with 64 cubes on a loop, with some straight pieces and some bend pieces — these are generally impossible to solve by hand. An example with 125 pieces, all bends, has proven to be very hard to solve by hand. A correspondent sent a different solution from the one that came with the puzzle.

One can also ask for multiple forms. For example, a sequence of 8, with all B's, can be made into a 2×4 rectangle or a $2 \times 2 \times 2$ cube and this is the only sequence of 8 which will do both. There is no sequence of 12 which will make a 2×6 and a 3×4 rectangle. The $2 \times 3 \times 3$ block requires $6 \leq b \leq 10$ and some of its solutions will make a 2×6 or a 3×4 or neither. When $b = 10$ there is only one way to make the 2×6 but many ways to make the $2 \times 2 \times 3$. Sequences with 16 or 24 start giving so many possibilities that they have not been examined. For the 3×9 rectangle, it is easy to see that our original sequence will not work. There are examples with $4 \leq b \leq 17$ but it is not clear how to show $b \geq 18$ is impossible.

Abel *et al.* [1] show that the problem of deciding whether it can be done at all is NP-complete, meaning that the time required to do so for an $N \times N \times N$ cube may grow more quickly than any polynomial function of N. (This is not surprising since the underlying Hamiltonian Path problem is also NP-complete; a discussion of this is beyond the scope of this volume.) Knowing that there is a solution, however, would not necessarily tell how to perform the folding — but would motivate you, on recreational grounds, to try to find one!

Another question is to generalize the dimensions. The one-dimensional problems are trivial as only straight pieces can be used. Two-dimensional problems are already non-trivial. The $2 \times n$ rectangle ($n \geq 2$) can be done with all B's, with only 2 B's or with any intermediate number of B's, i.e., for $2 \leq b \leq n$. The 3×3 can only be done with $b = 4$ or 5 and the 3×4 can only be done with $4 \leq b \leq 8$. The $2 \times m \times n$ can be done with all B's, or with $b = 4m - 2$ (taking $m \leq n$). It is not known that all intermediate values of b are possible.

The path puzzles have given rise to mathematical research into "bent" hamiltonian paths and cycles in any dimension, where every connection is a turn [2], and into which N^2 paths can be folded into a flat square (whether this problem is NP-complete is unknown [1]). This turns out to be related to what are known as combinatorial Gram codes. In which successive objects or positions differ in a minimal way; another chapter would be needed to follow this idea. (A recreational example is change-ringing of church bells, in which the order in which bells are rung can change only in specified ways.)

Bibliography

[1] Z. Abel, E. D. Demaine, M. L. Demaine, S. Eisenstat, J. Lynch and T. B. Schardl. "Finding a Hamiltonian Path in a Cube with Specified Turns is Hard." *Information and Media Technologies*, 8 (2013) 685–694.

[2] F. Ruskey and J. Sawada, "Bent Hamilton Cycles in Grid Graphs." *Electronic Journal of Combinatorics*, 10 (2003).

[3] jaapsch.net/puzzles/snakecube.

[4] cantaforda.com/cfcl/eryk/puzzles/chain_cube.

Chapter 6

Recurring Binomial Coefficients

Pascal's triangle has been a source of problems for centuries. Everyone who has been shown some hidden pattern in that famous array of numbers, wonders if they can find some new pattern? Probably not without some effort as this is a well-plowed field. However, we have found a new problem to explore in this chapter. If you look, you will find that repetitions of numbers are rare and any given entry is repeated very few times. Recall the entries of Pascal's triangle are the binomial coefficients, so we will just use the term binomial coefficient.[1]

Let $N(a)$ be the number of times a occurs as a binomial coefficient $\binom{n}{k}$. Clearly $N(1)$ is infinite and $N(a)$ is finite for all $a > 1$. Small values are easy to calculate: $N(2) = 1, N(3) = N(4) = N(5) = 2$, $N(6) = 3$. We will establish that $N(a) = O(\log a)$.[2] We conjecture that $N(a)$ is never larger than some fixed constant bound for all $a > 1$. Erdös, in a private communication, agreed with this conjecture but thinks it must be very hard. Later he suggested trying to show $N(a) = O(\log \log a)$.

Proposition 6.1. $N(a) = O(\log a)$.

---

[1]We assume some basic familiarity with these numbers. For example $\binom{n}{k} = \frac{n!}{k!(n-k)!}$. Further, we use the "big-oh" notation, where $N(a) = O(\log a)$ is short-hand for $N(a) < c \log a$, for some constant c (for large a).

[2]This result first appeared in the *American Mathematical Monthly*, 78(4) (April 1971), 385–386. This was supported by the Italian National Research Council.

Proof.   Let b be the least integer such that $\binom{2b}{b} > a$. Now

$$\binom{i+j}{i} = \binom{i+j}{j}$$

is monotonically increasing in i and in j; hence

$$\binom{b+i+b+j}{b+i} \geq \binom{b+b+j}{b} \geq \binom{2b}{b} \geq a$$

for all $i, j \geq 0$. Thus $\binom{i+j}{j} = a$ implies $i < b$ or $j < b$.

Again by monotonicity, for each value of i (or j),

$$\binom{i+j}{j} = a$$

has at most one solution. Hence $N(a) < 2b$. Now $\binom{2b}{b} \geq 2^b$ so we have

$$a \geq \binom{2(b-1)}{b-1} \geq 2^{b-1};$$

hence $b \leq \log_2 a + 1$, and $N(a) \leq 2 + 2\log_2 a = O(\log a)$.   □

6.1   Recurring Binomial Coefficients and Fibonacci Numbers

In this section, it is shown that there are infinitely many solutions to the equation

$$\binom{n+1}{k+1} = \binom{n}{k+2}.$$

given by $n = F_{2i+2}F_{2i+3} - 1$, $k = F_{2i}F_{2i+3} - 1$, where F_n is the nth Fibonacci number, beginning with $F_0 = 0$. So infinitely many binomial coefficients occur at least six times. The method and results of a computer search for repeated binomial coefficients, up to 2^{48}, will be given.

In the previous section, it was conjectured that the number of times an integer can occur as a binomial coefficient is bounded.

A computer search up to 2^{48} has revealed only the following seven nontrivial repetitions:

$$120 = \binom{16}{2} = \binom{10}{3};$$

$$210 = \binom{21}{2} = \binom{10}{4};$$

$$1540 = \binom{56}{2} = \binom{22}{3};$$

$$7140 = \binom{120}{2} = \binom{36}{3};$$

$$11628 = \binom{153}{2} = \binom{19}{5};$$

$$24310 = \binom{221}{2} = \binom{17}{8}$$

$$\text{and } 3003 = \binom{78}{2} = \binom{15}{5} = \binom{14}{5}.$$

It is known [2] that the only numbers which are both triangular, i.e., $= \binom{n}{2}$ for some n, and tetrahedral, i.e., $= \binom{n}{3}$ for some n, are 1, 10, 120, 1540 and 7140. The first two are trivial and the last three were also found by the computer, giving a check on the search procedure.

The coefficient 3003 occurs in the following striking pattern in Pascal's triangle:

$$
\begin{array}{cccccc}
1001 & & 2002 & & 3003 \\
& 3003 & & 5005 & \\
& & 8008 & &
\end{array}
$$

Some years ago it was discovered that this gives the only solution to

$$\binom{n}{k} : \binom{n}{k+1} : \binom{n}{k+2} = 1 : 2 : 3, \tag{6.1}$$

and that there is at most one solution to this relation when the right-hand side is replaced by $a:b:c$. This led to determining solutions when the right-hand side was $a:b:a+b$, or, equivalently and more simply, solutions of

$$\binom{n+1}{k+1} = \binom{n}{k+2}. \tag{6.1A}$$

Solution

From Eq. (6.1A), we have $(n+1)(k+2) = (n-k)(n-k-1)$. Set $m = n+1$, $j = k+2$, thus obtaining $m^2 + (1-3j)m + j^2 - j = 0$.

Solving for m gives

$$m = \frac{-1 + 3j \pm \sqrt{5j^2 - 2j + 1}}{2}.$$

For this to make sense, we must have that $5j^2 - 2j + 1$ is a perfect square, say v^2. We can rewrite this as

$$(5j - 1)^2 - 5v^2 = -4. \tag{6.2}$$

Letting $u = 5j - 1, C = -4$, we have

$$u^2 - 5v^2 = C. \tag{6.3}$$

This can be completely solved by standard techniques [5, Section 58, p. 204ff]. The basic solutions are: $9 \pm 4\sqrt{5}$ when $C = 1$; $2 \pm \sqrt{5}$ when $C = -1$; and $1 \pm \sqrt{5}$ and $4 \pm 2\sqrt{5}$ when $C = -4$. The class of solutions determined by $4 + 2\sqrt{5}$ is the same as the class determined by $4 - 2\sqrt{5}$, i.e., the class is ambiguous, in the terminology of [5]. Hence all solutions are given by

$$u_i + v_i\sqrt{5} = (-1 + \sqrt{5})(9 + 4\sqrt{5})^i, \; u_i + v_i\sqrt{5} = (1 + \sqrt{5})(9 + 4\sqrt{5})^i,$$

$$u_i + v_i\sqrt{5} = (4 + 2\sqrt{5})(9 + 4\sqrt{5})^i,$$

and their conjugates and negatives.

Let $F_0 = 0$, $F_1 = 1$, $F_{n+1} = F_n + F_{n-1}$ define the Fibonacci numbers and let $L_0 = 2, L_1 = 1, L_{n+1} = L_n + L_{n-1}$ define the Lucas numbers.

Lemma 6.1. $(L_n + F_n\sqrt{5})(9 + 4\sqrt{5}) = L_{n+6} + F_{n+6}\sqrt{5}$.

Proof. Let $\alpha = \frac{1+\sqrt{5}}{2}$, $\beta = \frac{1-\sqrt{5}}{2}$. It is well known that $F_n = (\alpha^n - \beta^n)/\sqrt{5}$ (the Binet formula), and $L_n = \alpha^n + \beta^n$, and so $L_n + F_n\sqrt{5} = 2\alpha^n$. Hence the lemma reduces to showing $\alpha^6 = 9 + 4\sqrt{5}$, which is readily done. $\qquad\square$

Since the basic solutions $u_0 + v_0\sqrt{5}$ given above are respectively

$$L_{-1} + F_{-1}\sqrt{5}, \quad L_1 + F_1\sqrt{5}, \text{ and } L_3 + F_3\sqrt{5},$$

the general solution of Eq. (6.3) can be written as

$$L_{2i-1} + F_{2i-1}\sqrt{5}, \text{ for } i = 0, 1, \ldots,$$

and we may now ignore conjugates and negatives.

To solve Eq. (6.2), we must have $5j - 1 = L_{2i-1}$. From the Binet formula, one may obtain $L_i \equiv 2 \cdot 3^i \pmod 5$ and hence $L_i \equiv -1 \pmod 5$ if and only if $i \equiv 3 \pmod 4$. Recalling that $j = k + 2 \geq 2$, the solutions of Eq. (6.2) are thus

$$j = (L_{4i+3} + 1)/5, \quad v = F_{4i+3}, \quad i = 1, 2, \ldots.$$

By standard manipulations, we obtain

$$j = F_{2i}F_{2i+3} + 1, \quad k = F_{2i}F_{2i+3} - 1,$$

$$m = F_{2i+2}F_{2i+3} = \frac{L_{4i+5} - 1}{5}, \quad n = F_{2i+2}F_{2i+3} - 1. \tag{6.4}$$

Finally, observe that

$$\binom{n}{k} : \binom{n}{k+1} = (k+1) : (n-k) = F_{2i} : F_{2i+1}$$

hence

$$\binom{n}{k} : \binom{n}{k+1} : \binom{n}{k+2} = F_{2i} : F_{2i+1} : F_{2i+2}.$$

The case $i = 1$ gives $n = 14$, $k = 4$, and

$$\binom{15}{5} = \binom{14}{6} = 3003.$$

The case $i = 2$ gives $n = 103$, $k = 38$, $k + 2 = 40$, and

$$\binom{104}{39} = \binom{103}{40} = 6121\,81827\,43304\,70189\,14314\,82520.$$

This number does not occur again as a binomial coefficient. The next values of (n, k) are $(713, 271)$ and $(4894, 1868)$.

Remarks

Equation (6.1A) has also been solved by Lind [4]. Hoggatt and Lind [3] have dealt with some related inequalities.

The coefficients

$$N = \binom{n+1}{k+1} = \binom{n}{k+1} = \binom{N}{1}$$

give us infinitely many binomial coefficients occurring at least six times. This has also been noted in [1, Theorem 3]. Since 3003 happens to be also a triangular number, one might hope that some more of these values might also be triangular. However $\binom{103}{40}$ is not triangular and does not occur as any other binomial coefficient. These determinations are described below. Other patterns in the repetitions found have not been discerned.

One might try to extend the pattern of Eq. (6.1A) and try to find

$$\binom{n}{k+4} = \binom{n+1}{k+3} = \binom{n+2}{k+2}.$$

This would require two solutions of Eq. (6.1A) with consecutive values of n and inspection of Eq. 6.4 shows this is impossible.

Lemma 6.1 is a special case of the general assertion that the solutions u_i, v_i of

$$u_i + v_i\sqrt{D} = (u_0 + v_0\sqrt{D})(a + b\sqrt{D})^i$$

both satisfy the same second-order recurrence relation:

$$u_{n+1} = 2au_n + (b^2 D - a^2)u_{n-1}.$$

(In our particular case: $F_{n+6} = 18F_n - F_{n-6}$.) It is unclear if the fact that the three basic solutions happen to neatly fit together into a single linear recurrence is a happy accident or a general phenomenon.

Above we considered Eq. (6.2). More generally,

$$C(n, k-1) : C(n, k) : C(n, k+1) = a : b : c$$

has at most one solution, given by

$$n + 1 = \frac{(a+b)(b+c)}{b^2 - ac}; \quad k = \frac{a(b+c)}{b^2 - ac}.$$

No criterion is known on a, b, c which makes n and k integral and which is simpler than equations above. Even though n and k are integral if and only if $(b^2 - ac)/GCD(a, b)$ divides $b + c$, this is not any simpler than seen above. This analysis is similar to the Ass and Mule problem (see the companion volume). In fact, it is possible for just one of n and k to be integral — in contrast to the Ass and Mule Problem. Consider $a, b, c = 1, 5, 5$ (or 1, 5, 7) and 8, 10, 11. It is also clear that one must have $b^2 - ac > 0$, but this is hardly sufficient — consider 1, 3, 4. Similar questions can be asked for three consecutive binomial coefficients along a diagonal: $C(n-1, k):C(n, k):C(n+1, k) = a{:}b{:}c$, but here we have $a < b < c$, so some questions are a bit different. This topic is discussed in [10].

6.2 Computer Search

If we let $M(k)$ be the first integer a such that $N(a) = k$, we have: $M(1) = 2, M(2) = 3, M(3) = 6, M(4) = 10, M(6) = 120$. We have found that $M(8) = 3003$ and this is the only solution to $N(a) \geq 8$ with $a \leq 2^{23}$. Further values would be interesting to know.

Two separate computer searches were made. First an ALGOL program was used to search up to 2^{23} on the London Polytechnics' ICL 1905E. All the 4717 binomial coefficients $\binom{n}{k}$ with $k \geq 2, n \geq 2k$ and less than 2^{23} were formed by addition and stored in rows corresponding to the diagonals of Pascal's triangle. As each new coefficient was created, it was compared with the elements in the preceding rows. Since each row is in increasing order, a simple binary search was done in each preceding row and the process is quite quick. All the repeated values given in the Introduction were already determined in this search.

The second search was carried out using a FORTRAN program on the University of London Computer Centre's CDC 6600. Although the 6600 has a 60-bit word, it is difficult to use integers bigger than 2^{48} and overflow occurs with such integers. Consequently, we were only able to search up to 2^{48}. There are about 24×10^6 triangular numbers and about 12×10^4 tetrahedral numbers up to this limit. It is impractical to store all of these, so the program had to be modified. Fortunately, the results of [2], mentioned above, implied that we did not have to compare these two sets. A subroutine was written to determine if an integer N was triangular or tetrahedral. This estimates the J such that $J(J+1)/2 = N$ by $J = \lfloor \sqrt{2N} \rfloor - 1$ and then computes the successive triangular numbers until they equal or exceed N. Two problems of overflow arose, Firstly, if N is large, the calculation of the first triangular number to be considered, i.e.,

$J(J+1)/2$, may cause an overflow when $J(J+1)$ is formed. This was resolved by examining $J(\bmod 2)$ and computing either $(J/2)(J+1)$ or $J((J+1)/2)$. Secondly, if N is larger than the largest triangular number less than 2^{48}, the calculation of the successive triangular numbers will produce an overflow before the comparison with N reveals we have gone far enough. This was resolved by testing the index of the triangular numbers to see if overflow was about to occur. The test for tetrahedral numbers was similar, but requires testing $J(\bmod 6)$.

The search then proceeded much as before. All coefficients $\binom{n}{k}$ with $k \geq 4$ and $n \geq 2k$ and less than 2^{48} were formed by addition and stored in rows. As each coefficient was formed, the subroutine was used to see if it was triangular or tetrahedral and binary search was used to see if it occurred in a preceding row.

It is startling that the second search produced no new results. The results 210, 11628, 24310, and 3003 were checked, which gave us some confidence in the process. The program was rerun with output of the searching steps and this indicated that the program works correctly. Perhaps someone can extend this to higher limits and see if there are more repetitions.

The calculation of $N = \binom{103}{40}$ and the computational determination that it was not triangular were also complicated by overflow, since $N > 2^{48}$. First we attempted to compute only the 103rd row of the Pascal triangle by use of $\binom{103}{k} = \frac{104-k}{k}\binom{103}{k-1}$, using double precision arithmetic. However, this showed inaccuracies in the units place, beginning with $k = 33$. We then computed the entire triangle up to the 103rd row $(\bmod\ 10^{14})$ by addition. We could then overlap the two results to get N. The double precision calculation had been accurate to 27 of the 29 places.

We applied the idea of the subroutine to determine if N were triangular. This required some adjustments. Since $2N$ is bigger than 2^{96}, one cannot truncate $\sqrt{2N}$ to an integer. Instead $\sqrt{N/2}$ was calculated, truncated to an integer and then doubled. Then the process of the subroutine was carried out, working in double precision real form. N was found to lie about halfway between two consecutive triangular numbers. These results for N were independently checked by Cecil Kaplinsky using multiprecision arithmetic on an IBM 360.

In a personal letter, D. H. Lehmer pointed out that one could determine that N was not triangular by noting its residue (mod 13). Following up on this suggestion, we computed the Pascal triangle (mod p) for small primes. Since $\binom{n}{k}$(mod p) is periodic as a function of n [8, 9] and [6] (Theorem 38), one can deduce that $N \not\equiv \binom{n}{k}$ for various ks by examination of N(mod p) and the possible values of $\binom{n}{k}$(mod p). For example, $N \equiv 4$ (mod 13), but $\binom{n}{k} \neq 4$ (mod 13) for $k = 2, 4, 6, 7, 8, 9, 10, 11, 12$. Using the primes 13, 19, 29, 31, 37, 53, 59 and 61, one can exclude all possibilities for k, other than 39 and 40 and hence N occurs exactly six times.

On the basis of the computer search and the scarcity of solutions of Eq. 6.1A, we are tempted to make the following assertion.

Conjecture 6.1. *No binomial coefficient is repeated more than* 10 *times.*

Stoll [7] does not add any repetitions beyond those given here.

Bibliography

[1]   H. L. Abbott, P. Erdös and D. Hanson. "On the Number of Times an Integer Occurs as a Binomial Coefficient." *American Mathematical Monthly*, 81 (1974) 256–261.

[2]   E. T. Avanesov. "Solution of a Problem on Figurate Numbers." *Acta Arithmetica* 12 (1966/67) 409–420 (Russian) [See the review: MR 35, No. 6619.]

[3]   V. E. Hoggatt, Jr. and D. A. Lind. "The Heights of Fibonacci Polynomials and an Associated Function." *The Fibonacci Quarterly*, 5, 2 (April 1967) 141–152.

[4]   D. A. Lind. "The Quadratic Field $Q(\sqrt{5})$ and a Certain Diophantine Equation." *The Fibonacci Quarterly*, 6(1) (February 1968) 86–93.

[5]   T. Nagell. *Introduction to Number Theory.* 2nd ed., Chelsea Publ. Co., 1964.

[6]   D. Singmaster. "Divisibility of Binomial and Multinomial Coefficients by Primes and Prime Powers." In *A Collection of Manuscripts Related to the Fibonacci Sequence*, V. E. Hoggatt, Jr. & M. Bicknell-Johnson, eds., The Fibonacci Association, 1980, 98–113.

[7]   M. Stoll. "How to Solve a Diophantine Equation." In *An Invitation to Mathematics*, D. Schleicher & M. Lackman, eds., Springer-Verlag, 2011, 9–19.

[8]   W. F. Trench, "On Periodicities of Certain Sequences of Residues." *American Mathematical Monthly*, 67 (1960) 652–656.

[9]   S. Zabek, "Sur la périodicite modulo m des suites de nombres $\binom{n}{k}$." *Ann. Univ. Mariae Curie-Sklodowska* Sect. A, 10 (1956) 37–47.

[10] M. Chamberland, *Single Digits In Praise of Small Numbers*, Princeton University Press. Princeton & Oxford, 2015, 2nd ptg & 1st PB ptg, 2017, Repetition in Pascal's Triangle, pp. 196–197.

Chapter 7

Sums of Squares and Pyramidal Numbers

One of the delights of mathematics is discovering different routes to the same result. It displays the nature of unseen connections that some people find as evidence that mathematics is "discovered" rather than "invented". Many find searching for new proofs of known results simultaneously fun and sublime. Finding ways to prove identities leads to pleasant and surprising insights.

This chapter discusses the well-known formula for the sum of the first n squares:

$$S(n) = 1^2 + 2^2 + 3^2 + \cdots + n^2 = \frac{n(n+1)(2n+1)}{6}. \qquad (7.1)$$

Ian Anderson gave five ways to prove this [1]. His methods are mostly algebraic. I wondered if there were more geometric ways of obtaining the formula. Several methods were found and are presented here. These methods are based on combinatorics, counting balls in piles, and are suitable as an exploratory activity for students.[1]

First, we recall that another well-known identity:

$$1 + 3 + 6 + \cdots + \binom{n+1}{2} = \binom{n+2}{3}. \qquad (7.2)$$

This tells us that a triangular pyramid or tetrahedron of cannonballs, with n balls on an edge, has $\binom{n+2}{3}$ balls altogether and it is a

---

[1]This appeared in *The Mathematical Gazette*, 66(436) (June 1982), 100–104.

consequence of the additivity property of Pascal's triangle. Because of this, we call $\binom{n+2}{3}$ the nth pyramidal (or tetrahedral) number. Since $\binom{n+1}{2}$, the nth triangular number, is usually denoted $T(n)$, we denote $\binom{n+2}{2}$ by $P(n)$.

The following methods continue Anderson's numbering. This discussion is self-contained so the reader need not start with Anderson's paper. Briefly, Anderson works with $\binom{n+2}{3} + \binom{n+1}{3}$ since it is algebraically equal to $\frac{1}{6}n(n+1)(2n+1)$. His first four methods are algebraic methods, like telescoping series. The fifth method is a novel mapping from subsquares of an $(n+1) \times (n+1)$ square integer lattice.

Method 6

Consider a square pyramid with n levels. The number of balls is then $S(n) = 1^2 + 2^2 + \cdots + n^2$. Divide the pyramid into two parts by a vertical plane parallel to a diagonal of the square and displaced by half a ball from the diagonal. Then we get two triangular pyramids with n and $n-1$ levels having $1 + 3 + 6 + \cdots + \binom{n+1}{2} = P(n)$ and $0 + 1 + 3 + \cdots + \binom{n}{2} = P(n-1)$ balls, respectively, as in Figure 7.1. Thus $S(n) = P(n) + P(n-1)$ which yields Eq. (7.1). (This is a geometric visualization of Anderson's third method.)

Method 7

Consider the same square pyramid and let the vertical plane pass through the diagonal. This plane then contains $1 + 2 + 3 + \cdots + n = T(n)$

Figure 7.1.   A pyramid of balls.

balls and the rest of the pyramid forms two triangular pyramids of $n - 1$ levels. Thus

$$S(n) = T(n) + 2P(n - 1) = \binom{n + 1}{2} + 2\binom{n + 1}{3}$$

$$= \binom{n + 2}{3} + \binom{n + 1}{3},$$

which again gives the result in (7.1). It may be worth observing that the vertical planes parallel to the diagonal contain $1, 3, 6, \ldots$ balls, i.e., triangular numbers of balls.

Method 8

Consider a cube of n^3 balls, with coordinates $(1, 1, 1), (1, 1, 2),$ $\ldots, (1, 1, n), \ldots, (n, n, n)$. We can extract a square pyramid with its apex at $(1, 1, 1)$ and its base on the face $x = n$. From the remainder of the cube, we can extract two square pyramids of $n - 1$ levels with apices at $(1, 2, 1)$ and $(1, 2, 2)$ and bases on the faces $y = n$ and $z = n$, respectively. There remains a triangle of balls with vertices at $(1, 1, 2), (1, 1, n), (n - 1, 1, n)$. Thus

$$n^3 = S(n) + 2S(n - 1) + T(n - 1) = S(n) + 2S(n) - 2n^2 + T(n - 1),$$

which again yields the result of (1).

Method 9

Consider the same cube of n^3 balls. This contains three square n-level pyramids with apex $(1, 1, 1)$ and bases on the faces $x = n, y = n$ and $z = n$, respectively. These overlap in three n-level triangles which all overlap along the main diagonal. Thus

$$n^3 = 3S(n) - 3T(n) + n,$$

which again yields the desired result. (This method gives a nice illustration of the inclusion–exclusion principle.)

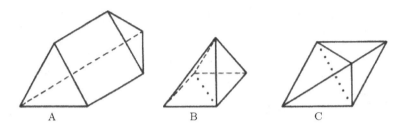

Figure 7.2.   A bisection of a pyramid of balls.

Method 10

This method shows that $S(n) = \frac{1}{4}\binom{2n+2}{3}$, i.e., $\frac{1}{4}P(2n)$, a result which Anderson could not find directly. Consider a triangular pyramid of $2n$ levels. We will divide it into congruent quarters by the following method.

Consider two opposite edges of a regular tetrahedron. Their common perpendicular is the line joining their midpoints, which we call a *midline* of the tetrahedron. The three midlines are mutually perpendicular and all pass through the centroid of the tetrahedron. A plane passing through two of these lines will be called a *midplane*. A midplane meets the tetrahedron in a square cross-section, with edge equal to half the edge of the tetrahedron. A midplane bisects the tetrahedron into two congruent pentahedra. (This bisection is a fairly well-known puzzle. It is surprising how difficult it is for some people to form a pyramid from two such pieces.) See Figure 7.2.

Now repeat the bisection process with another midplane. We get four congruent pieces! Each piece is a pyramid on a rhomboidal base having a 60° angle. That is, the base consists of two equilateral triangles. The apex of the pyramid is over the center of one of these triangles and forms a regular tetrahedron with that triangle. Figure 7.2B shows this pyramid and Figure 7.2C shows a top view of it. The tetrahedral parts of the pyramids are located at the four vertices of our original tetrahedron and the rhomboidal bases are on the four faces, as shown in top view in Figure 7.3. (This quadrisection is a little known and quite difficult puzzle.)

A little examination shows that each of our rhomboidal pyramids contains $S(n)$ balls when the original tetrahedron has $2n$ balls on each edge. Thus $4S(n) = P(2n)$, as desired.

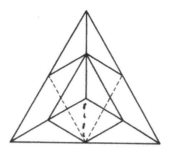

Figure 7.3. A quadrisection of a pyramid of balls.

While working on the above, a number of related questions come to mind, which are left as exercises.

- What happens when you bisect the tetrahedron by all three midplanes at once?
- Use similar reasoning to deduce $1 \cdot n + 2(n-1) + 3(n-2) + \cdots + n \cdot 1 = P(n)$.
- Consider n^3 as three square pyramids of $n-1$ levels plus some remainder in order to derive Eq. (7.1).
- Relate method 9 to a proof that a square pyramid has volume one-third of the base times the height. (Trisection of a cube in this way also makes a nice puzzle.)
- The number $P(1)+P(2)+P(3)+\cdots+P(n) = \binom{n+3}{4}$ is the volume of a four-dimensional simplex of balls with n along each edge, and the idea extends to higher dimensions. Extend all the above results to higher dimensions. (You may find it easier to extend the algebraic versions first.)

Bibliography

[1]   I. Anderson. "Sums of Squares and Binomial Coefficients." *Mathematical Gazette*, 65(432) (June 1981) 87–92.

Chapter 8

The Bridges of Königsberg

In memorium Jeremy Wyndham.[1]

Königsberg, now Kaliningrad, in the Kaliningrad Oblast of the Russian Federation, was the capital of East Prussia and a German city until World War II (Figure 8.1). It is a few miles up the River Pregel [Pregolya in Russian] from the Baltic Sea. The Oblast (=region) is disconnected from the rest of Russia and found between modern Poland, Lithuania and the Baltic Sea.

The city was founded in 1255 and was the headquarters of the Teutonic Knights. It was a major German port and then Russia's only warm-water port to the Baltic. The (Herzog Albrecht or Albertina or Albertus) University of Königsberg was founded by Duke Albrecht of Prussia in 1544 and was one of the major universities in Germany. Immanuel Kant (1724–1804) was born, lived, taught and died here. At some time, the river was realigned and so there are two arms entering the city and joining there. At the confluence is a large island, Kneiphof. As the city grew during the middle ages, bridges were built, with the first in 1286. In 1542, a seventh bridge was built and this configuration remained until the first railway bridge was built in 1865.

---

[1]My late colleague Jeremy Wyndham was interested in the seven bridges problem and made inquiries which turned up several maps and postcards of Königsberg [Geheimes Staatsarchiv Preussischer Kulturbesitz, Berlin, items XX. HA Staatsarchiv Königsberg, Kartensammlung F 10.706] and a list of all the bridges and dates of construction [8].

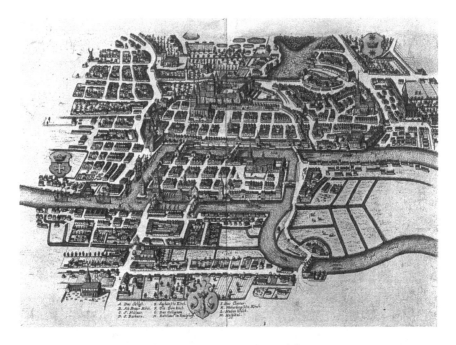

Figure 8.1.　Map of Königsberg, [7], c. 1641.

Apparently the citizens of Königsberg enjoyed walking about their city and attempted to make a walk which crossed all seven bridges, once each, but they were unable to find such a walk. In 1736, this problem came to the attention of the great Swiss mathematician, Leonhard Euler (1707–1783), then at the Imperial Academy of Sciences in St. Petersburg. (A 19th century article says Euler visited Königsberg, which is likely as it is between Berlin and St. Petersburg.) Euler wrote a note about it which easily showed that no such route is possible and this note is generally considered the first paper on "graph theory", one of the major branches of modern mathematics.

In view of the seminal importance of the problem of the seven bridges of Königsberg, it is surprising that very little has been done on the *history of the bridges* — there seems to be just one paper discussing the situation after the eighth bridge was built in 1865. My late colleague Jeremy Wyndham took an interest in this history about 1997 and obtained various maps, photographs and information

Figure 8.2. Holzbrücke, 1915 and Krämerbrücke, 1938.

from German sources. The photos in Figures 8.2 and 8.3 are three early 20th century postcards, with postmark dates.

These sources give a complete list of all the bridges and their dates of building and demolition up to WWII when all the bridges were destroyed — though the date of construction of bridge no. 4 is unclear. The 1997 Lonely Planet guide shows the present bridges.

In Graph Theory, a graph is quite different from the older and more common usages of the word. A graph is a configuration comprising a set of points or vertices, together with edges (undirected) or arcs (directed) joining them. The actual geometric layout is not considered; what is important is whether there is a connection or not. Some examples: a road map, where the points are the intersections and the edges are the roads; an electrical wiring diagram; the map of the London Underground; a family tree; the hierarchy of an organization; the points and edges of a polyhedron; the states and events in a project; an airline route map; etc. Problems of passing over every edge of a graph occur in numerous situations. Perhaps the most obvious is organizing postmen's routes (where each side of the road is viewed as an edge). Routes of snowplows, dustcarts, political canvassers, streetsweepers, meter readers, gutter/sewer/road/pavement/track/parking meter inspectors, farmers checking fences, etc. are all examples of such circuits. More abstract versions lead to techniques for getting out of labyrinths (Figure 8.3).

Though this problem is well known, to illustrate some new points we sketch Euler's original work. Euler's approach to the problem involved simplifying the map of Königsberg to four land areas (which can be considered as points), and seven bridges (which can be considered as edges), as in Figure 8.4.

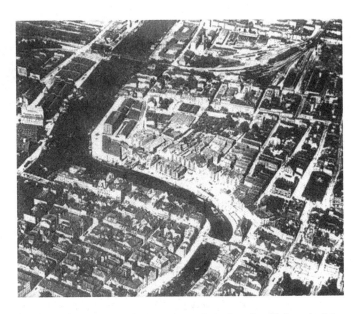

Figure 8.3.   Langgasser Brücke, Neue Eisenbahnbrücke, Krämerbrücke, c. 1925.

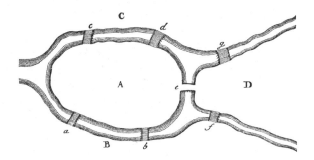

Figure 8.4.   Simplified map from Euler.

Each land area has a certain number of bridges joined to it: A has 5; B, C, D each have 3. Euler notes that any walk which crosses each bridge once must leave each area as often as it enters it, except for the beginning and ending areas, if these are not the same. He established this theorem, the first and perhaps best-known result in Graph Theory.

Theorem 8.1. *If such a walk exists, then every area, except possibly two, must have an even number of bridges connected to it.*

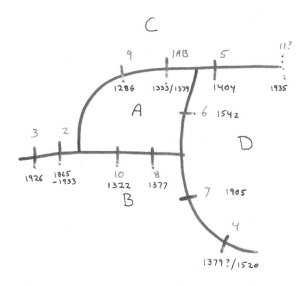

Figure 8.5.   Map with Bridges dated and labeled.

But in Königsberg, there are four areas with an odd number of bridges and hence there can be no such walk. Euler also observed that the sum of the bridge numbers (i.e., $5 + 3 + 3 + 3$) is twice the number of bridges (i.e., 14) and hence only an even number of these numbers can be odd.

Somewhat surprisingly, Euler did not give a proof of the converse result that if there are 0 or 2 areas with an odd number of bridges, then one can find a walk of the desired type and it was not until 1873 that a careful proof of this result was given [4]. In view of Euler's basic work, these walks are called Euler circuits (when they return to the starting point) or Euler paths (when the beginning and end may be different points).

A list of bridges is given in Figure 8.5, then a list of dates and a time chart, followed by an analysis of the number of Euler paths.

From the data, we can see that until the seventh bridge of 1542, there was always an Euler path. Up to 1542, the number of paths was always small enough to be computed by hand. In 1865, an eighth bridge was built and in 1876 a German mathematician named L. Saalschütz gave a lecture on the new situation [9] and stated there were now 384 Euler paths (where he counted each route twice, once from each end). With the ninth bridge of 1905, the number of paths gets so large that counting is tedious and error-prone, so we wrote a program to count all the paths. The 1865 data was my first test and was rather surprised to find the program gave the answer 416, a different answer than Saalschütz gave. He had missed two cases in his basic enumeration and these corresponded to 32 routes in the total. It seemed likely that no one had observed Saalschütz's error before.

There is a popular problem of counting Euler paths on the "envelope" pattern below. This has two forms and various books give different answers.

8.1   The Envelope Problem

It is a popular childhood puzzle to ask someone to draw a path over all the lines of the "envelope" pattern below, with each line covered just once. The endpoints of such a path can only be two of the possible corners. A puzzle book claimed that there are 50 different such paths.[2]

As an added complication, there are two ways to view the pattern. My initial view was that there were four points: 1, 2, 3, 4, joined by all six possible connections, with the extra connection E drawn from 3 to 4. In this view, the center point where lines B and F cross is not an actual connection and one cannot shift from line B to line F at this point and this is what the left-hand diagram is intended to convey. But the picture does have this point in it and perhaps one should consider the situation as shown in the right-hand diagram. I wrote a program to find all these paths and did not get 50 for either case (Figure 8.6).

An Eulerian path must have an even number of lines at each point, except possibly for the endpoints. In either of the given diagrams, we see that points 1 and 2 have three lines at them, while the other points have four lines, so points 1 and 2 must be the endpoints. For convenience, we just consider paths starting at 1; those starting at 2 will then be just the reversals of those starting at 1. (The diagrams

---

[2]I tried doing this by hand and did not get that number.

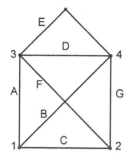 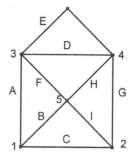

Figure 8.6. The Envelope Problem.

are also symmetric about the vertical midline and one could consider mirror image paths as being the same, which would divide our numbers by 2 since no path is equal to its reflection.)

In either case, the lines D and E are equivalent and interchangeable. So we will only count the paths where line D precedes line E and this again divides our problem by a factor of two. In the first situation, the pattern is also symmetric by interchanging A with B and F with G. We find the following 11 paths.

$$ADBCFEG, \; ADBCGEF, \; ADEFCBG, \; ADEFGBC,$$

$$ADGCBEF, \; ADGFEBC, \; AFCBDEG, \; AFGDEBC;$$

$$CFABDEG, \; CFDBAEG, \; CFDEABG.$$

Each of these gives four distinct paths by applying the two mentioned symmetries, so there are 44 paths from point 1 to point 2 (or 88 paths if we count both directions). This has now been checked by a computer program.

In the second situation, the counting is less easy and my original handwork omitted some cases which my program has identified. There are 22 paths starting with A, 22 starting with B and 16 starting with C, giving 60, each of which gives two distinct paths by interchanging D and E, so there are 120 paths from point 1 to point 2.

Sam Loyd had examined the eight bridges of Königsberg problem in his *Cyclopedia* of 1914 [5] (pp. 155 & 359–360) and found 416 routes, confirming my work. He also asked for the "shortest" route, without giving any distances Figure 8.7.

PROPOSITION---Tell just how many different routes there are, and which is the shortest.

The Bridges of Königsberg.
There are 416 ways of doing this trick of which the shortest route is via O-P, D-C, E-F, H-G, I-F, L-K, N-M and A-B, but as there are several million ways of not doing it, such a small matter as 416 routes may have been overlooked.

Figure 8.7. Loyd's version and his solution, 1914.

In the early 20th century, several more bridges were built and one was demolished, but there were always Euler paths until at least 1935. Then an eleventh bridge was built on a ring road, but it is so far out that it does not appear on any of the maps seen and is ignored.

8.2   The Pregel Bridges

1A. Dombrücke (Cathedral Bridge). Built between 1379. Demolished in 1379 when it was replaced by the Schmiedebrücke.

2. Alte Eisenbahnbrücke (Old Railway Bridge). Built in 1933, from the Ost- und Südbahnhof (East and South Rail Station) to the Altes Zollamt (Old Customs Office). In the dispute between the City and State railways in 1929, it was withdrawn from service [or confiscated] and shut down. Mühlpfordt [8] says it was demolished in 1933. However, Mallion [6] indicates that bridge 2 was still present in 1936, but gone in 1938.

3. Neue Eisenbahnbrücke (New Railway Bridge). Built in 1919–1926, from Hauptbhf. (Main Rail Station) parallel to Reichsstr. to Bhf. Holländer Baum. Swing bridge system. Used 16,000 m^3 of concrete. Opened in the presence of the Reichsbahn-präs. (President of the State Railway) Dr. Dorpmüller. [The 1930s map labels this Reichsbahnbrücke (State Railway Bridge) and says it is a rail and road bridge.]

4. Hohe Brücke (High Bridge) also known as Alte Brücke (Old Bridge). Certainly [or already] existing in 1500/1520. Built [presumably meaning rebuilt] in 1500–1520. [However, "zugestanden" seems to mean "granted" or "conceded", so this may mean that the bridge was authorized in 1377 and not built until 1500–1520.] Replaced in 1882.

5. Holzbrücke (Wood Bridge). Built in 1404. Renewed in 1901–1904 (with a commemorative plaque to the chronicler Herzog (Duke [this might possibly be a given name]) Albrechts Lukas David. Shown on the Lonely Planet map.

6. Honigbrücke (Honey Bridge). Completed in 1542. Named for the tubs of honey with which the so-called chief burgrave Besenrade bribed the Kneiphof councillors in the matter of overturning a Bierzeise [beer tax?], whereupon the angry citizens of the Altstadt (Old City) named the Kneiphofers "honey lickers". Replaced in 1882. Shown on the Lonely Planet map.

7. Kaiserbrücke. Opened in 1980s. Cost 540,000 Marks. Apparently destroyed in 1944. Mallion [6] (p. 31) says it was damaged in the war, was demolished in the 1980s and replaced by a facsimile foot bridge in 2005. Not on the Lonely Planet map, so presumably destroyed in 1944.

8. Köttelbrücke [Köttel is not an ordinary German word, but may be related to Kot, which can mean cot, cottage, shed; dirt, filth, dung, sewage, etc. The 1763 map labels it Kuttel Brücke and Kuttel means tripe or entrails.] Built in 1944? Replaced in 1886, and presumably destroyed in 1944.

9. Krämerbrücke (Shopkeeper's Bridge). Built in 1286 with small shops on it. = St. Georgsbrücke (St. George's Bridge) in 1339 [the meaning of the = is not clear to me, it may indicate a change of name or a rebuilding]. = Koggenbrücke [Koggen is not a normal German word, but may be related to Kog, Köge, reclaimed land; Koggenstr. runs a bit north and west of the bridge] in 1397 = officially Kokinbrücke [Kokin is not a normal German word] in 1548. Renewed, without shops, in 1787. Replaced in 1899/1900. Shown on the Lonely Planet map.

10. Langgasser Brücke (Long Lane Bridge) [Langgasse is the street which runs over this bridge, with prefix Vorstadt on the south side and Kneiphof on the island] = Grüne Brücke (Green Bridge). Built in 1322. Burnt in 1582, Replaced in 1590. This

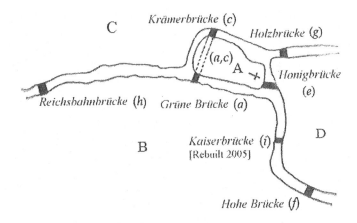

Figure 8.8.   From Mallion [6].

bridge, with its high winches [this sounds like it was a draw bridge?] remained in hand operation until its replacement in 1907. Shown on the Lonely Planet map. Mallion reports that bridges 9 and 10 have been combined into a single bridge — Estacada Bridge — which passes over the island, but there are pedestrian steps down to the island, so this does not change the connectivity.

11. Palmburger Brücke. Built in 1935 as a bypass route to Northeast Prussia. [Not shown on the maps, but the modern guide book shows a ring road to the east of the city, crossing the Pregel. What happens to the Alte Pregel — does it just peter out, or does it reconnect to the Neue Pregel?]

1B. Schmiedebrücke (Smith's Bridge). Built in 1379? after arbitration by Grand Master Winrichs allocated the cost equally to both cities. (Mallion reports two sources for the date of bridge 1B — 1379 and 1397, while Mühlpfordt gives 1379.) (It carried small shops. Renewed in 1787. Replaced in 1896. [Replaced Dombrücke. Not on the Lonely Planet map, so presumably destroyed in 1944.]

Supposedly all the bridges were destroyed in WW II. Since then at least six bridges have been built, but we have not yet obtained data on the dates of these. A 2006 map of the bridges is reproduced in Figure 8.8.

Analysis

We collect this data in a time chart in Figure 8.9.

With this data, we can determine just what graph was present at each time. We label the land regions with A, B, C, D in the

Date	Bridges	Number	Start	Number of paths/circuits
1286	010000	1	A	1 path
1322	110000	2	B	1 path
1333	120000	3	A	2 paths
1377	220000	4	A	8 circuits
			B	4 circuits
If bridge 4 built before bridge 8, we have:				
1377*	120010	4	A	2 paths
If both bridges 4 and 8 are built in 1377, we have:				
1377**	220010	5	B	4 paths
1404	220001	5	C	4 paths (isomorphic to 1377**)
If bridge 4 built before bridge 5, we have:				
1404*	220011	6	B	16 paths
1520	220011	6	B	16 paths (same as 1404*)
1542	221011	7		NONE
1865	221111	8	A	208 paths (Saalschütz got 192)
1905	221121	9	A	960 paths
1926	221221	10	A	4512 paths
1938	221121	9	A	960 paths (same as 1905)
Ignoring bridge 11, we have:				
1944	111111	6		NONE
If bridge 11 joins C and D, then we have:				
1935	221122	10		NONE
1944*	111112	7	A	44 paths
1980s	111111	6	A	NONE All areas have degree 3
the first envelope problem	111121	7	A	44 paths. Isomorphic to envelope prob.

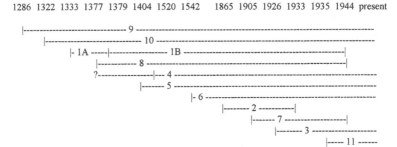

Figure 8.9.   Time Chart of the Pregel Bridges.

same way as Euler and Saalschütz, which places them schematically as in Fig. 8.4, where A is the island Kneiphof. Consult Figure 8.4. We can then describe the situation by listing the number of bridges between the regions A and B, A and C, A and D, B and C, B and D, C and D, *in that order*. For convenience, we express this as a six-digit number, followed by the total number of bridges. For example, the classic state of seven bridges can be denoted by 221011 and 7. My program computes the Eulerian paths or circuits in a graph and gives the number of paths in each case, from one given starting point.

From this analysis, we see that, during 1286–1935, there was just one period, of 323 years, 1542–1865, when no Euler path existed, while there are two periods, totaling 326 years, when Euler paths existed. Had Euler not lived in the right period, perhaps the problem would never have been studied and our postmen would be wandering about inefficiently.

From the analysis above, one can see that if a graph has $2k$ odd vertices, then it can be covered in k Euler paths, or in one path which repeats $k-1$ edges, or in a circuit which repeats k edges. In practice, this is a common situation and one then wishes to minimize the length or cost of the repeated edges. (This is known as the Chinese Postman Problem.)

8.3   Other Places

The idea of finding Euler circuits or paths is a fairly obvious problem. Euler gives an example with six areas, two islands and fifteen bridges, but only finds one circuit. Only one other reference has been seen to other places — namely [Coupy] who translates Euler into French and adds a note saying that "an interesting application of the problem of Euler can be made to Paris" but limits his region to that between the bridges of Iena and Austerlitz, with four areas and 24 bridges.

Looking at a recent map of Paris, finds 4 areas and 42 bridges. This reveals several difficulties in trying to work on other cities. (1) It is not clear how far to go. One would want an area which one could easily walk around. New York, London and Paris are too spread out. (2) For interest, one should have a limited number of islands. Venice is too complex and is not fully connected unless we permit connections by boat. New York has several main areas, but also has several small islands, e.g., Roosevelt Island, which has only one access, which raises the problem of connections which are only one-way. Perhaps some reader will take up this question and find other interesting examples.

Bibliography

[1]   B. and K. Cornwell. *Seven Bridges of Konigsberg*. International Film Bureau, 1965.

[2]   É. Coupy. "Solution d'un problème appartenant a la géométrie de situation, par Euler." *Nouv. Ann. Math.*, 10 (1851) 106–119. (Translation of Euler. Note on p. 119 applies it to the bridges of Paris.)

[3]   L. Euler. "Solutio problematis ad geometriam situs pertinentis." *Comm. Acad. Sci. Petropol.* 8 (1736 or 1741) 128–140. Also *Opera Omnia* 1, 7 (1923) 1–10. (Translation in Biggs, Lloyd and Wilson, 3–8, and in *Scientific American*, 189 (July 1953) 66–70.)

[4]   C. Hierholzer. "Ueber die Möglichkeit, einen Linienzug ohne Wiederholung und ohne Unterbrechung zu umfahren." *Math. Annalen*, 6 (1873) 30–32.

[5]   S. Loyd. "Problem of the bridges of Königsberg." pp. 155 & 359–360. (also *Mathematical Puzzles of Sam Loyd*, vol. 1, Martin Gardner, ed. Dover, 1959, problem 28, pp. 26–27 & 130–131.)

[6]   R. Mallion. "A contemporary Eulerian walk over the bridges of Kaliningrad." *British Society for the History of Mathematics Bulletin* 23(1) (2008) 24–36.

[7]   M. Merian the Elder. Engraved map of Königsberg.[3]

---

[3]I have a colored reproduction of this, dated as 1641. It has been attributed to M. Zeiller, *Topographia Prussiae et Pomerelliae*, c. 1650. Seen in a facsimile of

[8] H. M. Mühlpfordt. *Königsberg von A–Z.* 1972. [Geheimes Staatsarchiv Preussischer Kulturbesitz, Berlin, Bibliothek, item 17 M 92, 16–17.]

[9] L. Saalschütz. "[Report of a lecture.]" *Schriften der Physikalisch-Ökonomischen Gesellschaft zu Königsberg* 16 (1876) 23–24.

Cosmographica due to Merian in the volume on Brandenburg and Pomerania, see ZEILLER. There seem to be at least two versions of this picture.

Chapter 9

Triangles with Doubled Angles

In 1988, Tim Sole [9] asserted that in the 4, 5, 6 triangle, one angle was double another angle. This is the kind of observation that leads the mathematically minded to ask if there are more examples. This thinking led the Danish mathematician Mogens Larsen to explore this problem. His preliminary results appeared as "The Triangle with One Angle Twice Another". He showed that this leads to the relation $a^2 = b^2 + bc$ among the sides. In particular, he proved $\angle A = 2\angle B \leftrightarrow a^2 = b^2 + bc$. The preprint led to correspondence. And as often happens correspondence led to generalizations. What we both discovered are the results in this chapter.[1]

9.1 Geometry

The simplest examples of equivalent relations between angles and sides for a triangle $\triangle ABC$ with sides a, b, and c are well known, e.g.,

$$\angle A = \angle B \leftrightarrow a = b$$

and

$$\angle A = \angle B + \angle C \leftrightarrow a^2 = b^2 + c^2,$$

---

[1]Mogens Esrom Larsen, Københavns Universitets Matematiske Institut. The original is in *The Missouri Journal of Mathematical Sciences*, 3(3) (Fall 1991), 111–129. Larsen produced a slightly revised version in Danish, "Ækvivalente relationer mellem vinkler og sider i en trekant", which appeared in *Normat*, 41(4) (1993), 144–155. This is a slightly revised version of that.

because the angle relation is equivalent to $\angle A = \frac{\pi}{2}$. Schwering [6–8] and Heinrichs [3] (see also Dickson [2]) have studied relations of the form $\angle A = n\angle B$ and $\angle A = n\angle B + \angle C$ by the use of trigonometric functions and roots of unity. Willson [11] and Luthar [4] have considered the case $n = 2$, and recently Carroll and Yanosko [1] have generalized to the case of n rational. Maxwell [5] has considered triangles with $2\angle A = \angle B + \angle C$. In this chapter, we present elementary geometric proofs of the equivalences in each of these cases:

- Case A: $\angle A = 2\angle B$ $\quad\leftrightarrow\quad$ $a^2 = b^2 + bc$;
- Case B: $\angle A = 2\angle B + \angle C$ $\quad\leftrightarrow\quad$ $a^2 = b^2 + ac$;
- Case C: $2\angle A = \angle B + \angle C$ $\quad\leftrightarrow\quad$ $a^2 = b^2 + c^2 - bc$;
- Case D: $\angle A = 2(\angle B + \angle C)$ $\quad\leftrightarrow\quad$ $a^2 = b^2 + c^2 + bc$;
- Case E: $\angle A = 2(\angle B - \angle C)$ $\quad\leftrightarrow\quad$ $ba^2 = (b - c)(b + c)^2$.

Furthermore, we present the formulas for the complete set of integral solutions for each of these types of triangles.

Case A: All Triangles with $\angle A = 2\angle B$.

In triangle $\triangle ABC$ with $\angle A > \angle B$, we draw a line from A to D on a such that $\angle CAD = \angle B$ as in Figure 9.1. Then $\triangle ABC \sim \triangle DAC$, so that

$$\frac{b}{a} = \frac{d}{c} = \frac{a - x}{b}$$

or the two equalities

$$ad = bc$$

$$ax = a^2 - b^2 \tag{9.1}$$

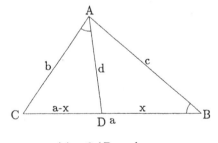

$$\angle A = 2\angle B \Leftrightarrow d = x$$

Figure 9.1.   Case A

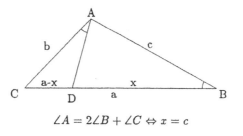

$$\angle A = 2\angle B + \angle C \leftrightarrow x = c$$

Figure 9.2.   Case B

from which we get the formula:

$$\frac{x}{d} = \frac{a^2 - b^2}{bc}.$$

Hence we conclude that $\angle A = 2\angle B$ if and only if the triangle $\triangle ABD$ is isosceles, or $x = d$, i.e.,

$$\angle A = 2\angle B \leftrightarrow \angle BAD = \angle B \leftrightarrow x = d \leftrightarrow a^2 - b^2 = bc.$$

This proves Case A.

Case B: All triangles with $\angle A = 2\angle B + \angle C$.
In a triangle with $\angle A > \angle B + \angle C$, we draw a line from A to D on a such that $\angle CAD = \angle B$ as in Figure 9.2. Again the triangles $\triangle ABC$ and $\triangle DAC$ are similar, so that we have Eq. (9.1).
  Now,

$$\angle A = 2\angle B + \angle C \leftrightarrow \angle A - \angle B = \angle B + \angle C.$$

But we have that

$$\angle BAD = \angle A - \angle B$$

and

$$\angle BDA = \angle B + \angle C.$$

Hence we have

$$\angle A = 2\angle B + \angle C \leftrightarrow \angle BAD = \angle BDA$$

$$\leftrightarrow x = c \leftrightarrow ac = ax \leftrightarrow ac = a^2 - b^2.$$

This proves Case B.

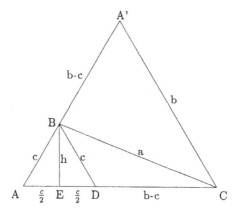

$$\angle A = \angle A' = \tfrac{\pi}{3} \text{ and } \angle BDC = \tfrac{2\pi}{3}$$

Figure 9.3.   Cases C & D

Cases C and D: All triangles with $\angle A = \tfrac{\pi}{3}$ or $\angle A = \tfrac{2\pi}{3}$.
The angle relations in Cases C and D are equivalent to the equations
$\angle A = \tfrac{\pi}{3}$ and $\angle A = \tfrac{2\pi}{3}$, respectively. The relations of the sides are
of course just the cosine relations for these angles. But a closer look
proves worthwhile. Suppose $\angle A = \tfrac{\pi}{3}$ and that $\angle C < \angle A < \angle B$. Signs
of equality gives the trivial case of equilateral triangles, $a = b = c$.
Then we draw two equilateral triangles with the angle $\angle A$ having
side lengths c and b as in Figure 9.3.

Then we notice the pair of solutions $\triangle ABC$ and $\triangle A'BC$ with
sides a, b, c and $a, b, b - c$, respectively. Furthermore, the triangle
$\triangle DBC$ has $\angle BDC = \tfrac{2\pi}{3}$ and sides $a, c, b - c$. So the solutions appear
three at a time. An elementary solution of the cosine relation comes
from Pythagoras applied to the triangles $\triangle ABE$ and $\triangle BCE$; i.e.,

$$c^2 - \left(\frac{c}{2}\right)^2 = h^2 = a^2 - \left(b - \frac{c}{2}\right)^2$$

hence

$$c^2 = a^2 - b^2 + bc$$

proving Case C from which

$$a^2 = c^2 + (b - c)^2 + c(b - c) \tag{9.2}$$

proving Case D.

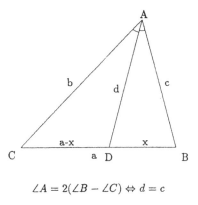

$$\angle A = 2(\angle B - \angle C) \Leftrightarrow d = c$$

Figure 9.4. Case E, beginning

Case E: All triangles with $\angle A = 2(\angle B - \angle C)$.
This equation is similar to the previous ones, so it is surprising that
it is equivalent to a third degree equation in the sides and it is much
harder to solve. We draw in a triangle, with $\angle C < \angle B$, the bisector
from A to D on a as in Figure 9.4.

Then $\angle ADB = \angle C + \frac{1}{2}\angle A$. Hence the relation is equivalent to
the triangle $\triangle DAB$ being isosceles.

$$\angle A = 2(\angle B - \angle C) \leftrightarrow \angle C + \frac{1}{2}\angle A = \angle B \leftrightarrow c = d.$$

But this time we need some further lines for support. We extend AB
over A to the point E, such that $AE = b$, and we extend AD over D
to the point F, chosen such that the angle $\angle DCF = \frac{1}{2}\angle A$, as shown
in Figure 9.5

The first extension gives us the isosceles triangle $\triangle CAE$, and
hence that $\angle ACE = \angle CEA = \frac{1}{2}\angle CAB$. Therefore

$$\triangle DAB \sim \triangle CEB.$$

The second extension gives us

$$\triangle DAB \sim \triangle CAF \sim \triangle DCF.$$

The angle relation is thus equivalent to the four similar triangles
being simultaneously isosceles.

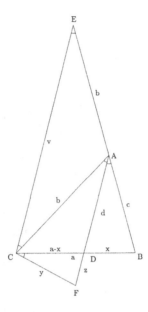

Figure 9.5.   Case E, conclusion

We now begin the proof of Case E. The four similar isosceles triangles and that $AD \parallel EC$, give two equal relations, namely

$$\frac{(a-x)}{b} = \frac{x}{c} = \frac{a}{b+c} = \frac{x}{y} = \frac{y}{d+z} \qquad (9.3)$$

and

$$\frac{x}{d} = \frac{z}{a-x}. \qquad (9.4)$$

From Eq. (9.3) we get

$$y^2 = z(d+z) = zd + z^2.$$

Using Eq. (9.4) we get

$$y^2 = x(a-x) + z^2.$$

Now using Eq. (9.3) again, this becomes

$$y^2 = \frac{c(a-x)^2}{b} + y^2 \left(\frac{a}{b+c}\right)^2$$

which can be rewritten as

$$y^2 \left(1 - \left(\frac{a}{b+c}\right)^2\right) = \frac{c(a-x)^2}{b}.$$

From this equation, we get the equivalence

$$y = a - x \leftrightarrow \frac{(b-c)}{b} = \left(\frac{a}{b+c}\right)^2,$$

which completes the proof of Case E.

Although this is considerably simpler than our original proof, we suspect a simpler proof can be found.

Geometric summary

It is striking that the identical figure arises in Cases A and B and a very similar figure arises in Case C. In Cases A and B, we considered two cases of equality between edges of $\triangle ABD$. The trivial case, $c = d$, gives $\angle C = 0$. Since $\triangle ABC \sim \triangle DAC$, an equality between sides of $\triangle DAC$ makes $\triangle ABC$ isosceles. We can make other simple constructions, e.g., find D so that $\angle ADB = \angle B$; but they do not lead to any other relations. In Case C, we have the same basic construction, but with AD bisecting $\angle A$. There are six possible equalities of sides in the triangles $\triangle ABD$ and $\triangle ACD$. The five other cases do not lead to any new relations. We have examined all linear relationships $\alpha \angle A + \beta \angle B + \gamma \angle C = 0$ with $\alpha, \beta, \gamma \in \{-2, -1, 0, 1, 2\}$ and all of these reduce to the five cases we have considered. Initially we thought all these cases would lead to second degree relations among the sides, so Case C and its difficulty were quite unexpected.

9.2 Diophantine Analysis

Recall by Diophantine analysis we mean the requirement of integral solutions. One of the inspirations for this study is the fact that the 4, 5, 6 triangle has one angle double another [7, 9, 11]. It is very remarkable that the smallest integral solutions of many of our cases have sides which are consecutive integers.

$(2, 3, 4) = $ (b, c, a) is a solution of Case B.

$(3, 4, 5) = $ (b, c, a) is a right triangle.

$(4, 5, 6) = $ (b, c, a) is a solution of Case A.

$(6, 7, 8) = $ (c, a, b) is a solution of Case E.

$(3, 5, 7) = $ (b, c, a) is a solution of Case D.

$(3, 7, 8) = $ (c, a, b) is a solution of Case C.

$(5, 7, 8) = $ (c, a, b) is a solution of Case C.

We might add $(1, 2, 3) = (a, b, c)$ is a solution of Case A and it is the smallest integral-sided scalene triangle. An obvious question is: what problem has (5, 6, 7) as an integral solution? We also know that (13, 14, 15) is the smallest Heronian triangle with consecutive sides [2, 10], so a general question is: what can be said about the integral-sided triangles $(b - 1, b, b + 1)$? We have not been able to make any progress on these questions.

All solutions to the Diophantine equations

Case A: The complete solution to $a^2 = b^2 + bc$ is

$$(a, b, c) = r(pq, p^2, q^2 - p^2) \text{ for } (q, p) \text{ coprime,}$$

$$2p > q > p, \ r \text{ arbitrary.} \tag{9.5}$$

Proof. If (a, b, c) are pairwise coprime, then Case A is written

$$a^2 = b(b + c).$$

Hence $b = p^2$ for some p and $b + c = q^2$ for some q, with $q > p$ and (q, p) coprime. The triangle inequality $a + b > c$ then gives $q^2 - pq - 2p < 0$ or $q < 2p$. Taking $q = 3, p = 2$ gives $a = 6, b = 4$ and $c = 5$. □

Case B: The complete solution to $a^2 = b^2 + ac$ is

$$(a, b, c) = r(q^2, pq, q^2 - p^2) \text{ for } (q > p) \text{ coprime, } r \text{ arbitrary.} \tag{9.6}$$

Proof. If (a, b, c) are pairwise coprime, then Case B is written

$$b^2 = a(a - c).$$

Hence $a = p^2$ for some p and $a - c = q^2$ for some q with $p > q$ coprime. Taking $p = 2, q = 1$ gives $a = 4, b = 2$ and $c = 3$. □

Cases C and D: The complete solutions to Cases C and D, respectively, are

$$(a, b, c) = \frac{r}{4}(3p^2 + q^2, \ 4pq, \ |3p^2 - q^2| - 2pq)$$

$$\text{and } (a, b, b + c) \text{ and } (a, b + c, c),$$

where (p, q) are coprime odd numbers satisfying $q < p$ or $3p < q$ and r is arbitrary.

Proof. Case C follows from Eq. (9.2). To solve Case D we assume (a, b, c) to be pairwise coprime and hence b or c is odd. By symmetry we may choose b as odd. Now we write Case D as

$$3b^2 = (2a + 2c + b)(2a - 2c - b). \tag{9.7}$$

A common prime factor s of these two factors would be a prime factor of $3b^2$ and therefore odd. It would also be a factor in their sum, $4a$, and hence in a, and in their difference, $4c + 2b$, and hence in c, contrary to assumption. So the two factors in Eq. (9.7) are coprime, and we must have $b = pq$ such that the following sets are equal:

$$\{2a + 2c + b, \ 2a - 2c - b\} = \{3p^2, \ q^2\}.$$

By addition, we get $4a = 3p^2 + q^2$ and by subtraction, we get

$$4c + 2b = |3p^2 - q^2|$$

from which the first statement follows. Note $p = 2$ and $q = 3$ give $(a, b, c) = (7, 3, 5)$, etc. In order to have $c > 0$, we must have either $q < p$ or $3p < q$. Then $p = 1$ and $q = 5$ give $(a, b, c) = (7, 5, 3)$, etc. □

Case E: The complete solution to Case E is

$$(a, \ b, \ c) = r(p(2q^2 - p^2), \ q^3, \ q(q^2 - p^2)) \tag{9.8}$$

for $p < q$ coprime integers and r arbitrary.

Proof.   A common factor of a and b will be a factor of c, so we may assume a and b to be coprime. Let p be a prime factor of b to the power n, i.e.,

$$b = p^n \cdot d, \tag{9.9}$$

such that p and d are coprime. Now let

$$c = p^m \cdot f, \tag{9.10}$$

where m is such that p and f are coprime. Then we substitute Eqs. (9.9) and (9.10) in Case E, getting

$$p^n d a^2 = p^{3m}(p^{n-m}d - f)(p^{n-m}d + f)^2.$$

Hence $n = 3m$. We conclude that there is a q such that $b = q^3$ and a g such that $c = q \cdot g$. Hence we have

$$q^3 a^2 = q^3(q^2 - g)(q^2 + g)^2.$$

Let $h = q^2 - g$. Then, from

$$a^2 = h(2q^2 - h)^2$$

we conclude that h is a square, say $h = p^2$. Then $a = p(2q^2 - p^2)$, $b = q^3$ and $c = q(q^2 - p^2)$. Now $c > 0$ requires $p < q$. For $p = 1$, $q = 2$, we get $(a, b, c) = (7, 8, 6)$. $\qquad\qquad\square$

Right angled and isosceles triangles

In Cases C and D, we obviously have isosceles solutions and even equilateral in Case C. In these cases, a right angle excludes integral solutions. If the triangles in Case A or B are isosceles, the only possibility will be

$$p^2 - q^2 = pq,$$

but this equation has no rational solutions, because

$$\frac{p}{q} = \frac{1 \pm \sqrt{5}}{2}.$$

If a triangle in Case E shall be isosceles, then $\angle A = \angle B$ or $\angle A = \angle C$. If $\angle A = \angle B$, then $2\angle C = \angle A = \angle B$ and we are in

Case A with $\angle B = \angle C$, proved impossible above. If $\angle A = \angle C$, then $a = c$ and hence

$$2pq^2 - p^3 = q^3 - qp^2$$

without integral solutions. If a triangle in Case A is right angled, then either $\angle B$ or $\angle C$ is a right angle. In the first case, the triangle becomes isosceles; in the second case, it becomes a 30°–60°–90° triangle; neither of these can be integral. The Case B can be rewritten as

$$\angle A = \frac{\pi}{2} + \frac{1}{2}\angle B$$

so these triangles are always obtuse-angled. In Case E $\angle C < \angle B$, so only $\angle B$ or $\angle A$ may be right. If $\angle B$ is right, then $\angle A = 0$, so this is not the case. If $\angle A$ is right, then $\angle B = \frac{3\pi}{8}$ and $\angle C = \frac{\pi}{8}$, so one triangle exists. But the sides must satisfy Case E and Pythagoras. Eliminating a^2 from the equation, we obtain

$$b(b^2 + c^2) = (b - c)(b + c)^2,$$

with the only solution

$$b = (\sqrt{2} + 1)c.$$

Integral areas

When an integer sided triangle has integral area, it is called Heronian. Of course, none of the triangles of Cases C or D can avoid a factor of $\sqrt{3}$, so these are less interesting. It is useful to rewrite the parameterization equations. (9.5), (9.6) and (9.8) as follows. Now p, q are coprime, with $p < q$.

	a	b	c
$\angle A = 2\angle B$	pq	p^2	$q^2 - p^2$
$\angle A = 2\angle B + \angle C$	q^2	pq	$q^2 - p^2$
$\angle A = 2(\angle B - \angle C)$	$2pq^2 - p^3$	q^3	$q^3 - qp^2$

These forms make it easy to give a useful table of possible sides:

p	q	p^2	pq	$q^2 - p^2$	q^2	$2pq^2 - p^3$	q^3	$q^3 - qp^2$
1	2	*1	2	3	4	7	8	6
1	3	*1	3	8	9	17	27	24
2	3	4	6	5	9	28	27	15
1	4	*1	4	15	16	31	64	60
3	4	9	12	7	16	69	64	28
1	5	*1	5	24	25	49	125	120
2	5	*4	10	21	25	92	125	105
3	5	9	15	16	25	123	125	80
4	5	16	20	9	25	136	125	45
1	6	*1	6	35	36	71	216	210
5	6	25	30	11	36	235	216	66
1	7	*1	7	48	49	97	343	336
2	7	*4	14	45	49	188	343	315
3	7	*9	21	40	49	267	343	252
4	7	16	28	33	49	328	343	231
5	7	25	35	24	49	365	343	168
6	7	36	42	13	49	372	343	91
1	8	*1	8	63	64	127	512	504
3	8	*9	24	55	64	357	512	440
5	8	25	40	39	64	515	512	312
7	8	49	56	15	64	553	512	120

The * means that a Case A triangle does not exist, because $q \geq 2p$.

Of course, it is not obvious whether any of these has integral area. It is convenient to make use of the area formula of Heron:

$$\Delta = \sqrt{s(s-a)(s-b)(s-c)},$$

where $s = \frac{1}{2}(a + b + c)$. In Case A, we obtain

$$s = \frac{1}{2}(pq + p^2 + q^2 - p^2) = \frac{1}{2}q(p+q),$$

$$s - a = \frac{1}{2}qp + \frac{1}{2}q^2 - pq = \frac{1}{2}q(q-p),$$

$$s - b = \frac{1}{2}qp + \frac{1}{2}q^2 - p^2 = \frac{1}{2}(q-p)(q+2p),$$

$$s - c = \frac{1}{2}qp + \frac{1}{2}q^2 - q^2 + p^2 = \frac{1}{2}(q+p)(2p-q).$$

So we get

$$\Delta^2 = \left(\frac{1}{4}\right)^2 \cdot q^2 \cdot (p+q)^2 \cdot (q-p)^2 \cdot (2p+q)(2p-q).$$

For Δ to be an integer, $(2p+q)(2p-q) = 4p^2 - q^2$ must be a square. Considerations (mod 4) show that q must be even, which makes Δ^2 an integer and also makes p odd. Any common factor of $2p+q$ and $2p-q$ must divide their sum $4p$ and their difference $2q$, but GCD $(4p, 2q)$ can only be 2 or 4. So, either

$$2p+q = 4s^2 \text{ and } 2p-q = 4t^2$$

or

$$2p+q = 2s^2 \text{ and } 2p-q = 2t^2.$$

So we have two possibilities,

$$p = s^2 + t^2 \text{ and } q = 2(s^2 - t^2) \tag{9.11}$$

or

$$p = \frac{1}{2}(s^2 + t^2) \text{ and } q = s^2 - t^2 \tag{9.12}$$

with Eq. (9.11) to apply for s, t coprime of different parity, and Eq. (9.12) for s, t coprime and both odd. The area then becomes either

$$\Delta = q \cdot (q^2 - p^2) \cdot s \cdot t \tag{9.13}$$

or

$$\Delta = \frac{1}{2} \cdot q \cdot (q^2 - p^2) \cdot s \cdot t, \tag{9.14}$$

where s, t have different parity in Eq. (9.13) and s, t are both odd in Eq. (9.14).

So we can make the following table of Heronian triangles.

				a	b	c	$\triangle$
s	t	p	q	pq	p^2	$q^2 - p^2$	$\frac{1}{2}qcst$
2	1	5	6	30	25	11	132
3	1	5	8	40	25	39	468
4	1	17	30	510	289	611	73320
5	1	13	24	312	169	407	24420
5	2	29	42	1218	841	923	387660
6	1	37	70	2590	1369	3531	1483020

In Case B, we obtain

$$s = \frac{1}{2}(q^2 + pq + q^2 - p^2) = \frac{1}{2}(q + p)(2q - p),$$

$$s - a = q^2 + \frac{1}{2}(pq - p^2) - q^2 = \frac{1}{2}p(q - p),$$

$$s - b = q^2 + \frac{1}{2}(pq - p^2) - pq = \frac{1}{2}(q - p)(2q + p),$$

$$s - c = q^2 + \frac{1}{2}(pq - p^2) - q^2 + p^2 = \frac{1}{2}p(q + p).$$

So we get

$$\Delta^2 = \left(\frac{1}{4}\right)^2 \cdot p^2 \cdot (q^2 - p^2)^2 \cdot (2q - p) \cdot (2q + p).$$

We can use Eqs. (9.11) and (9.12) with p and q interchanged. We get for the area

$$\Delta = p \cdot (q^2 - p^2) \cdot s \cdot t \quad (s, \ t \text{ even})$$

or

$$\Delta = \frac{1}{2} \cdot p \cdot (q^2 - p^2) \cdot s \cdot t \quad (s, \ t \text{ odd})$$

and the following table:

				a	b	c	$\triangle$
s	t	p	q	q^2	pq	$q^2 - p^2$	$\frac{1}{2}qcst$
3	2	10	13	169	130	69	4140
4	3	14	25	625	350	429	72072
5	3	16	17	289	272	33	3960
5	4	18	41	1681	738	1357	488520
6	5	22	61	3721	1342	3237	2136420
7	5	24	37	1369	888	793	333060
7	6	26	85	7225	2210	6549	7151508

In Case E, we obtain

$$s = \frac{1}{2}(2pq^2 - p^3 + q^3 + q^3 - qp^2) = \frac{1}{2}(q + p)(2q^2 - p^2),$$

$$s - a = pq^2 + q^3 - \frac{1}{2}p^3 - \frac{1}{2}qp^2 - 2pq^2 + p^3 = \frac{1}{2}(q - p)(2q^2 - p^2),$$

$$s - b = pq^2 + q^3 - \frac{1}{2}p^3 - \frac{1}{2}qp^2 - q^3 = \frac{1}{2}p(q - p)(2q + p),$$

$$s - c = pq^2 + q^3 - \frac{1}{2}p^3 - \frac{1}{2}qp^2 - q^3 + qp^2 = \frac{1}{2}p(q + p)(2q - p).$$

So we get

$$\Delta^2 = \left(\frac{1}{4}\right)^2 \cdot p^2 \cdot (q^2 - p^2)^2 \cdot (2q^2 - p^2)^2 \cdot (2q + p) \cdot (2q - p).$$

We can again use Eqs. (9.11) and (9.12) with p and q interchanged. We get for the area the formulas

$$\Delta = p \cdot (q^2 - p^2) \cdot (2q^2 - p^2) \cdot s \cdot t \quad (s, t \text{ even})$$

or

$$\Delta = \frac{1}{2} \cdot p \cdot (q^2 - p^2) \cdot (2q^2 - p^2) \cdot s \cdot t \quad (s, t \text{ odd})$$

and the following table:

s	t	p	q	a $2pq^2 - p^3$	b q^3	c $q^3 - qp^2$	$\triangle$ $\frac{1}{2q}cast$
3	2	10	13	2380	2197	897	985320
4	3	14	25	14756	15625	10725	75963888
5	3	16	17	5152	4913	561	1275120
5	4	18	41	54684	68921	55637	1484123760
7	5	24	37	51888	50653	29341	720075720

Conclusion

It is striking that the common feature of the problems investigated is the isosceles triangles involved, see Figure 9.6. Some line segment from A to D on $a = CD$ is drawn, and then either $x = d, d = c, c = x$ or even $d = c = x$. Because of this similarity, the interpretations as angle relations are very similar. But in spite of this, the side relations are different and, in particular, the Case E with $d = c$ is surprisingly complicated. Nevertheless, the question of possible Heronian triangles is answered by the application of the very same Diophantine equation in the three solvable cases,

$$4p^2 - q^2 = r^2$$

proving a sort of similarity between the side relations also.

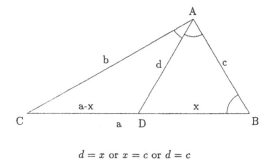

$d = x$ or $x = c$ or $d = c$

Figure 9.6.

Bibliography

[1] J. E. Carroll and K. Yanosko. "The Determination of a Class of Primitive Integral Triangles." *Fibonacci Quarterly,* 29 (1991) 3–6.

[2] L. E. Dickson. "Triangles, Quadrilaterals and Tetrahedra with Rational Sides: Rational or Heron Triangles." in *History of The Theory of Numbers,* Volume II: Diophantine Analysis. G. E. Stechert & Co, 1934 (also Chelsea Publ. 1952) Chapter V: 191–201. (Also "Triangles with Rational Sides and a Linear Relation Between the Angles," 213–214.)

[3] J. Heinrichs. "Aufgabe: Dreiecke mit ganzzahligen Seiten anzugeben so dass $\alpha = n\beta + \gamma$ wird." *Zeitschr. math. u. naturwiss. Unterricht,* 42 (1911) 148–153.

[4] R. S. Luthar. "Integer-sided Triangles with One Angle Twice Another." *College Mathematics Journal,* 15 (1984) 55–56.

[5] E. A. Maxwell. "Triangles Whose Angles are in Arithmetic Progression." *Mathematical Gazette,* 42 (1958) 113–114.

[6] K. Schwering. "Über Dreiecke, in denen ein Winkel das Vielfache eines andern ist." *Jahresbericht über das Königliche Gymnasium Nepomucenianum zu Coesfeld im Schuljahre 1885–86,* 85 (1886) 3–7.

[7] K. Schwering. *100 Aufgaben aus der niederen Geometrie nebst vollständigen Lösungen: Aufgabe 56: Von einem Dreieck ist gegeben: die Grundlinie BC, der zugehörige Höhenfusspunkt D und die Bestimmung $\alpha = 2\beta$.* Herdersche Verlagshandlung, 1891, 88–89.

[8] K. Schwering. "Ganzzahlige Dreiecke mit Winkelbeziehungen." *Archiv der Mathematik und Physik* (3) 21 (1913) 129–136.

[9] T. Sole. *The Ticket to Heaven and Other Superior Puzzles.* Penguin, London, 1988, 70 & 79.

[10] G. Wain and W. W. Willson, "13, 14, 15: An Investigation." *Mathematical Gazette,* 71(455) (1987) 32–37.

[11] W. W. Willson. "A Generalisation of a Property of the 4, 5, 6 Triangle." *Mathematical Gazette,* 60(412) (1976) 130–131.

Chapter 10

Quasicrystals and the University

All members of London South Bank University have probably noticed that the previous South Bank Polytechnic's coat of arms featured an unusual geometric pattern of a regular pentagon surrounded by five other regular pentagons (Figure 10.1).[1]

In 1973, I invited Roger Penrose, FRS, to speak at our Department seminar. Penrose was then Rouse Ball Professor of Mathematics at Oxford, famous for his invention of the "Impossible Triangle" used by Escher and one of the iconic images of the 20th century. More recently, he is Sir Roger, author of *The Emperor's New Mind* and *Shadows of the Mind*, and has been Gresham Professor of Geometry. He was fascinated by the Polytechnic letterhead and contemplated how he could use pentagons to fill the plane, a problem which he had worked on previously. In 1974, he discovered methods of filling in the gaps with other shapes in systematic but non-periodic patterns, such as shown in Figure 10.2.

Such patterns had been investigated before, but were complex (e.g., the first example, in 1966, used about 20,000 different shapes of tiles) and Penrose had tried to find simpler ones when the Polytechnic letterhead led him to look at pentagonal patterns and discover his pattern which was much simpler than previous ones. Penrose continued thinking about such patterns and evolved two pieces (which

---

[1]This was written in 1978. I revised it in 1992 and an abridged form appeared as part 2 of *Arms and the University* in *South Bank News*, Autumn 1994, 30. I have updated it.

Figure 10.1.   The present coat of arms and the former coat of arms.

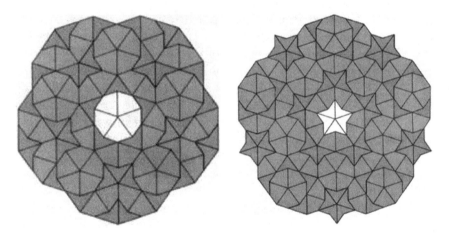

Figure 10.2.   Two tilings with the Penrose tiles.

he patented), called a kite and a dart, which can be assembled, following certain rules, into uncountably many packings of the plane. Remarkably, there is no way to use these pieces to fill the plane periodically.

Since then, the "Penrose Tiles" have fascinated many mathematicians, physicists and especially crystallographers as the packings exhibit a kind of five-fold symmetry which cannot occur in normal crystals. "The crystallographic restriction", which is one of the first results about crystals, says five-fold symmetry cannot occur. The Penrose packings and their solid analogues are called "quasicrystals". Mathematicians have shown that they can be viewed as projections of periodic tilings in higher dimensional space, along directions with irrational slope. More practically, if one does the Fourier analysis of the X-ray scattering from a quasicrystal, the result shows five- or ten-fold symmetry. In 1984, Daniel Shechtman observed such a scattering pattern from a real alloy of aluminium and manganese, now called Shechtmanite, but his group delayed announcing this for three years as they didn't believe it.

The appearance of this result in November 1984 completely disrupted a conference on crystallography the next January — "The whole conference turned into a study of that paper" recalls one of the organizers. Over 100 quasicrystalline alloys have since been found. Indeed a number of earlier occurrences have been discovered where the pictures had been discarded because the ten-fold symmetry displayed was thought to be an error. The quasicrystals are a totally unexpected new form of matter. In his 1994 Presidential Address to the Royal Society, Sir Michael Atiyah discussed the Penrose Pieces as an example of how mathematics can still produce novel results of physical significance. The behavior of quasicrystals is not yet understood and recent work indicates they have quite a different structure than the Penrose tilings, but there is some hope that they may produce stronger and lighter metals, so that this flight of fancy may lead to real flights of aircraft. Atiyah reported that they were being used as a non-scratch coating for French frying pans. Penrose (1990) says they are an example of a macroscopic effect of quantum mechanical laws. Quasicrystals have forced crystallographers to radically redefine what they think a crystal is — it is now "anything whose X-ray diffraction pattern has sharp, bright spots"!

Searching for quasicrystals on the internet reveals tens of thousands of references — there have been at least seven international conferences and about a score of books. A variation of the pieces, called "Perplexing Poultry" has been marketed as a kind of high powered jigsaw puzzle. The Mathematical Institute in Oxford has a display of the pieces. Mathematical institutions in the US and Australia have used the pattern to tile courtyards and hallways. Somewhat more amusingly, Kimberly Clark, makers of Kleenex toilet paper, used Penrose's pattern as a way of bulking "Quilted" toilet paper, but they failed to consult Penrose who has a copyright on the pattern and brought suit against Kimberly-Clark. This appears to have been settled out of court and the offending product has not been seen since then. A later form of the tiles has been used to tile a patio at Penrose's college, Wadham College, Oxford.

Roger Penrose was awarded the Order of Merit in 2000 and half of the Nobel Prize in Physics in 2020. In October 2011, Daniel Shechtman was awarded the Nobel Prize in Chemistry.

Any bibliography must be inadequate. Here are some of the earliest references and a few general articles.

Bibliography

[1] R. Penrose. "The Role of Aesthetics in Pure and Applied Mathematical Research." *Bull. Inst. Math. Appl.*, 10 (1974) 266–272.

[2] M. Gardner. "Penrose Tiling." *Penrose Tiles to Trapdoor Ciphers*, Freeman, 1989, 1–29. Column extended to two chapters.

[3] R. Penrose. "Pentaplexity." *Eureka*, 39 (1978) 16–22. (Also *Mathematical Intelligencer*, 2 (1979) 32–37).

[4] D. S. Shechtman, I. Blech, D. Gratias and J. W. Cohn. "Metallic Phase with Long Range Orientational Order and No Translational Symmetry." *Physical Review Letters*, 53(20) (12 November 1984) 1951–1953.

[5] D. R. Nelson. "Quasicrystals." *Scientific American*, 255, 2 (August 1986) 32–41 & 112.

[6] P. J. Steinhardt. "Quasicrystals." *American Scientist*, 74 (November/December 1986) 586–596.

[7] R. Penrose. *The Emperor's New Mind.* Oxford University Press, 1989, 132–138 & 434–437.

[8] M. W. Browne. "'Impossible' Form of Matter Takes Spotlight in Study of Solids." *New York Times*, (5 September 1989) 17 & 20.

[9] R. Penrose. *Shadows of the Mind.* Oxford University Press, 1994, 29–33 & 62.

[10] I. Stewart. "Bathroom Tiling to Drive You Mad." *New Scientist*, (24 September 1994) 14.

[11] M. Atiyah. "Anniversary Address by the President." *Supplement to Royal Society News*, 7(12) (November 1994).

[12] M. Senechal. *Quasicrystals and Geometry.* Cambridge University Press, 1995.

[13] J. Kay. "Top Prof Goes Potty at Loo Roll 'Rip-Off'." *The Sun*, (11 April 1997) 7.

[14] P. McGowan. "It Could End in Tears as Maths Boffin Sues Kleenex Over Design." *The Evening Standard*, (11 April 1997) 5. Also "Kleenex art that ended in tears." *The Independent*, (12 April 1997) 2. And "For a knight on the tiles." *Independent on Sunday*, (13 April 1997) 24.

[15] D. Trull. "Toilet Paper Plagiarism." *Parascope*, 1997. (Online journal).

[16] B. A. Cipra. "A gem of a definition." *SIAM News* 31(1) (January/February 1998).

[17] M. W. Browne. "New Data Help Explain Crystals that Defy Nature." *New York Times*, (24 November 1998) F–1.

[18] A. Boyle. "From Quasicrystals to Kleenex." *Plus* 16 (September 2001). (Online journal).

Chapter 11

The Wobbler

The Wobbler is a geometric device invented by Frederick Flowerday[1] some years ago. It consists of two interlocked circles. The planes of the circles are orthogonal and the centers of the circles are separated by $\sqrt{2}$ times the common radius. In practice, it is made from two circular discs, each with a radial notch. When the Wobbler is rolled on a plane, it executes a most fascinating wobbly motion. Observation indicates that the center of gravity remains at, or nearly at, constant height. We demonstrate that it does remain at constant height and that this holds if and only if the separation of centers is $\sqrt{2}$ times the radius. We also study the distance between the points of contact and show that this is constant if and only if the separation of centers is equal to the radius. Such a device was invented by Paul Schatz [1].[2]

If it is not easy to visualize the geometry then consider a huge version, called "Rocking Toy" that was made by the artist Anthea Alley, in Figure 11.1. The holes are for artistic reasons only.[3]

11.1 The Height of the Center of Gravity

Let the discs have radius r and let the separation of centers be $2d$ as shown in Figure 11.2. Clearly the center of gravity, C, will be d

---

[1]Flowerday, who contributed to this chapter, died in 2003.

[2]The original of this chapter appeared in *Eureka* 50 (1990) 74–78.

[3]Made in 1969, this uses 48" diameter disks, 1" thick, with holes about 12" diameter in the middle.

Figure 11.1.

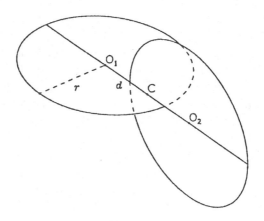

Figure 11.2.

from each of the centers which are denoted O_1 and O_2. There are two natural positions of the Wobbler. In Position 1, each disc is at 45° to the base plane. An end view is shown in Figure 11.3(a). Then the height h_1 is $\frac{r}{\sqrt{2}}$ and the relative height η_1 is $\frac{1}{\sqrt{2}}$.

In Position 2, one disc, say the second, is perpendicular to the base plane, as in Figure 11.3(b). Let A_1, A_2 be the points of contact. Then $A_1C = r+d$, $A_1O_2 = r+2d$, $O_2A_2 = r$, so that $\eta_2 = \frac{h_2}{r} = \frac{r+d}{r+2d}$. Thus $\eta_1 = \eta_2$ if and only if $\frac{r+d}{r+2d} = \frac{1}{\sqrt{2}}$, which gives us $\delta = \frac{d}{r} = \frac{1}{\sqrt{2}}$ or $2\delta = \frac{2d}{r} = \sqrt{2}$ as the relative separation of the centers.

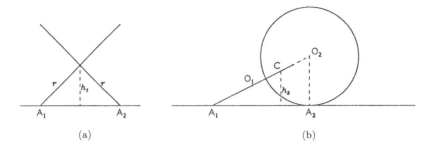

Figure 11.3. Position 1 and Position 2.

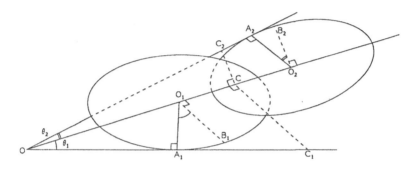

Figure 11.4.

This shows that $2\delta = \sqrt{2}$ is necessary for the center of gravity to have constant height, but this only considers the two special positions. The general position requires some care. As in Position 2, let the circles touch the base plane at A_1 and A_2. The intersection of the plane of disc 1 with the base plane is the tangent line to circle 1 at A_1. The tangent lines at A_1 and A_2 intersect at a point O in the base plane. The line of centers O_1O_2 also passes through O, which can be viewed as the intersection of the planes of the two discs and the base plane.

This is shown in Figure 11.4 where the line of centers OO_1O_2 is above the base plane which contains O, A_1 and A_2. Then OO_1A_1 and OO_2A_2 are right triangles. Let $\theta_i = \angle O_iOA_i$. Draw O_iB_i in the plane of disc i and perpendicular to OO_1O_2. Then $\angle OO_iB_i = 90°$, hence $\angle A_iO_iB_i = \theta_i$ and so θ_i is the angular measure of A_i from the natural point B_i which is where the Wobbler touches the base plane in Position 1.

Now draw axes CC_1, CC_2 through C and parallel to O_1B_1 and O_2B_2. Then CO, CC_1, CC_2 are cartesian axes based at C. We have $OO_1 = r \csc \theta_1$, so $OO_2 = OO_1 + O_1O_2 = r \csc \theta_1 + 2d$, which gives the basic relation:

$$\csc \theta_2 = \csc \theta_1 + 2\delta. \tag{11.1}$$

[Finding this was the hardest part of our analysis.]

We have $CC_i = CO \tan \theta_i$, so the equation of the base plane with respect to the CO, CC_1, CC_2 coordinates is

$$\frac{x}{CO} + \frac{y}{CO \tan \theta_1} + \frac{z}{CO \tan \theta_2} = 1.$$

Using the standard formula for the distance of a point to a plane, we have that the distance of C from the base plane, i.e., the height of the center of gravity, is

$$h = \frac{CO}{\sqrt{1 + \cot^2 \theta_1 + \cot^2 \theta_2}} = \frac{r \csc \theta_1 + d}{\sqrt{\cot^2 \theta_1 + \csc^2 \theta_2}},$$

so the relative height η is

$$\eta = \frac{h}{r} = \frac{\csc \theta_1 + \delta}{\sqrt{\cot^2 \theta_1 + (\csc \theta_1 + 2\delta)^2}}$$

$$= \frac{\csc \theta_1 + \delta}{\sqrt{2}\sqrt{\csc^2 \theta_1 + 2\delta \csc \theta_1 + 2\delta^2 - \frac{1}{2}}}.$$

The expression in the second radical is

$$(\csc \theta_1 + \delta)^2 + \delta^2 - \frac{1}{2}.$$

Hence the relative height η is constant, at $\frac{1}{\sqrt{2}}$ if and only if $\delta^2 = \frac{1}{2}$, i.e., $\delta = \frac{1}{\sqrt{2}}$ or $2\delta = \sqrt{2}$. This establishes our first result — the generalized Wobbler has center of gravity at constant height if and only if $2\delta = \sqrt{2}$.

Physically, this implies that the Wobbler is (meta)stable in any position. We have actually verified this by showing that the center of gravity lies over the line A_1A_2 joining the two points of contact, if and only if $2\delta = \sqrt{2}$. The point C', which is in the base plane directly under C, has coordinates $\frac{r^2}{2CO}(1, \cot\theta_1, \cot\theta_2)$, while

$$A_1 = (d + r\sin\theta_1, r\cos\theta_1, 0) \text{ and } A_2 = (-d + r\sin\theta_2, 0, r\cos\theta_2).$$

Then $A_1 - A_2 = t(C' - A_2)$ holds if and only if $\delta = \frac{1}{\sqrt{2}}$ and then $t = 2 + 2\delta\sin\theta_1$.

11.2 The Distance Between Contacts

In the last paragraph, we gave the coordinates of the contact points A_1 and A_2 with respect to the cartesian system based at C. Let $L^2 = (A_1 A_2)^2$ and $\frac{L}{r} = \lambda$.

We have

$$\lambda^2 = (2\delta + \sin\theta_1 - \sin\theta_2)^2 + \cos^2\theta_1 + \cos^2\theta_2$$

$$= 2 + 4\delta^2 + 4\delta\sin\theta_1 - 4\delta\sin\theta_2 - 2\sin\theta_1\sin\theta_2. \quad (11.2)$$

In Position 1, $\theta_1 = \theta_2 = 90°$, so $\lambda_1^2 = 2 + 4\delta^2$. In Position 2, $\theta_1 = 90°$, $\sin\theta_2 = \frac{r}{r+2d} = \frac{1}{1+2\delta}$, giving $\lambda_2^2 = 4\delta^2 + 4\delta$. Thus $\lambda_1 = \lambda_2$ if and only if $2\delta = 1$, i.e., the centers are separated by the radius, or each center lies on the other circle. When $\delta = \frac{1}{2}$, Eq. (11.1) yields $\sin\theta_1 = \sin\theta_2 + \sin\theta_1\sin\theta_2$. Using this and $\delta = \frac{1}{2}$ in Eq. (11.2) gives $\lambda^2 = 3$. This shows that λ is constant if and only if $\delta = \frac{1}{2}$. As we have seen, this does not give a constant height of center of gravity and so this device will not roll with the surprising ease of the Wobbler.

A version by Paul Schatz rolling down a plane is seen in [2].

11.3 Some Problems

We are primarily interested in the Wobbler, i.e., the device with $2\delta = \sqrt{2}$, but many of the following problems can be considered for the generalized Wobbler. We have studied the practical problems of the effect of the physical thickness of the discs and the effect of rounding the edges of the discs and have found adequate solutions for small thickness. We let $2t$ be the thickness and $\tau = \frac{t}{r}$ be the relative thickness. However, we have had no success with the following problems.

- As the Wobbler rolls on the base plane, the contact points A_i trace curves which appear somewhat like cycloids, or possibly epicycloids. The curve traced by the point O also seems interesting. What are these curves?
- Observation shows that the center of gravity C does not move in a straight line, but oscillates. What is this curve?
  This problem shows that the first problem is a bit harder than one might think. The oscillation of C shows that there is some frictional effect between the Wobbler and the base plane. Obviously, if there were no friction, the Wobbler would slide along. It seems that we can assume the Wobbler rolls without slipping, but this gets out of our knowledge and we are not clear about this. If so, then at a given position of the Wobbler, as in Figure 11.4, the arc lengths of the curves can be taken as $s_i = r\theta_i$. One can find all the relevant angles and lengths in the figure, but we cannot see how to relate them to coordinates in the base plane. If this can be done, then one can also consider the next problem. (See 9 for relevant work.)
- How does the axis of centers OO_1O_2 move as the Wobbler moves?
- If we connect all the corresponding points A_1 and A_2, we enclose the convex hull K of the Wobbler, shown in Figure 11.5. What are the volume and the surface area of K? By looking at one eighth of K and estimating the volume inside, we find that the volume is at least $3.2r^3$. This shape is discussed further below in the section on the Oloid.
- We can join the two circles with lines to produce a doubly-twisted surface as in Figure 11.5(b). This should be the minimal surface generated by the two circles, i.e., the soap film, if done correctly.

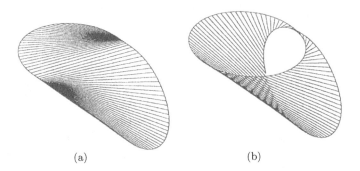

(a)   (b)

Figure 11.5.   Convex hull and possible soap bubble frame.

Can it be done correctly? If so, describe the surface and find its area.

John Watts, from South Bank Polytechnic, has made some progress, though his results are complicated. He lets β be the angle of inclination of the axis OO_1CO_2 and lets ψ be the angle of the projection of the axis on the x, y plane, with respect to some fixed direction. He finds

$$d\psi = \frac{\sin\beta(\cos^2\beta + 2)d\beta}{\cos\beta\sqrt{\cos^4\beta - 4\sin^2\beta}}$$

and the distance moved by C is given by

$$ds = rd\beta\frac{\cos^2\beta + 4\sin^2\beta + \sin^2\beta\cos^2\beta}{\cos^3\beta\sqrt{\cos^4\beta - 4\sin^2\beta}}.$$

If we take the initial position as having $\sin\beta = -(\sqrt{2} - 1)$ and with the axis along the y-axis, then $dx = ds\cos\psi, dy = -ds\sin\psi$. These formulas theoretically determine the curve of C, etc. He computes the surface area in D as 13.9236 and the volume as 3.2818, assuming $r = 1$.

11.4　Paul Schatz's Oloid

Paul Schatz (1898–1979) was a Swiss engineer and designer who was fascinated by wobbling motions. This led him to create a rotating ring of six tetrahedra, which form a cubical outline at one stage (Rick Flowerday had a similar Hexyflex). From this, he observed that the parts executed a complex three-dimensional wobbling motion and developed a way of supporting a container in the middle of a similar linkage as a device for mixing, e.g., for paint, powders, chemicals, pharmaceuticals, etc., or for polishing small parts, e.g., for watches. This turned out to be much more effective than previous mixing devices and several versions are still made, e.g., Turbula and Inversina. See [7, pp. 89–93 & 164–171], which shows examples which handle large drums.

In the late 1960s, he developed a simplified mixer utilizing the Oloid [1]. See Figure 11.6 for the internal structure. The shell of the Oloid is the convex hull of the two circles. Note that full circles are

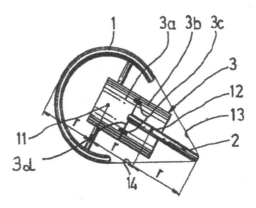

Figure 11.6.

not needed, only the section lying between the two tangents from the opposite end of the body is needed, as shown in the figure. If we put in two cross rods on the diameters of the two partial circles and connect these by a third rod, we get a simple version of the two-circle device. The third rod can be expanded into a substantial cylinder, as shown. One could make this hollow and put material into it, but Schatz suggests making a two part cover for the convex hull and putting the material into that. He then puts the Oloid on a sloping moving belt, so that the belt moves the Oloid up as it rolls down, thus staying in the same place in space, see Figure 11.7 Though almost any two circle device could be used, Schatz's patent prefers the Oloid configuration, which is not in the same proportions as the Wobbler. Perhaps the Wobbler configuration might require less energy, though it might not mix the region around the center of gravity as well.

In the early 1970s, Schatz developed another usage of the Oloid as a "wobble body" suspended by universal joints from rotating shafts in a rather complex manner which not only can be used for mixing material inside the body, but when suspended in fluid produces substantial mixing in the fluid, which has been used for aerating polluted water, etc. He took out patents in at least eight countries for this idea, but the Swiss patent has additional features, an extra page of drawings and some differences in one of the drawings. See [7] (pp. 124–127 & 172–178).

Schatz wrote a book expositing his ideas in 1975 and a second edition [7] with considerable updating appeared in 1998. It is covered

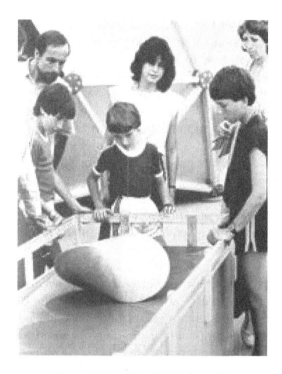

Figure 11.7. The Oloid, from [2].

on pp. 113–127, with additional material by Christian Ostheimer on pp. 143–148. The geometry is not clearly drawn in their Figure 167, but eventually one can see that his ω is $90° + \theta_1$ and his ω' is $90° - \theta_2$. Noting that $2\delta = 1$ for the Oloid, our Eq. (11.1), above is equivalent to Ostheimer's equation (1).

11.5 Other Results

Related work on the Wobbler continues to turn up. A. T. Stewart in Kingston, ON, had discovered it in the early 1960s and written about it in 1966, calling it a "Two-circle Roller" [3]. The article is only one page.

The object was also described and analyzed by Christian Ucke, an expert on physics toys in the Physics Department at the Technical University of Munich with some colleagues [4]. He gives a short description, mentioning Flowerday and citing Stewart and Schatz.

They say it is sold in Germany and Switzerland as Wobbler or Go-On and show a version using slotted plastic discs called Rondi. They give instructions on how to make one.

Ucke and Schlichting [5] gave a basic description and derivation of the basic result, though it hard to understand. They show this also works for elliptical discs! They attempt to find the path of the center of gravity. Our paper was not known to them when they wrote their paper.

Ucke has a recent illustrated exposition [6] which covers several variants of the idea. For example, take a disc and cut it in half and twist one half by 90°. Hiscox [8] gave this idea much earlier. He mounts the pieces on an axle at 45° to each of the semicircle's planes — in practice this requires some support between the pieces. Hiscox asserts that this rotates in a wind from any direction, except perpendicular to the axle, so it can be used as a stationary windmill, giving 60 days more work per year than conventional windmills. Ucke [6, p. 3] inadvertently asserts the shape of the twisted half circles has the property that its center of gravity has constant height, but he gives the spacing rule for circular discs and also for ellipses: the distance between centers is $\sqrt{4a^2 - 2b^2}$. He shows several examples of two-circle roller toys, including the Finnish Ensihammas, the version made from the German RONDI pieces and Flowerday's convex hull (Germans call this the "Torso", which means a developable surface).

The German designer Alexander Schenk adapted the convex hull into a salt and pepper shaker! He cuts the shape across the "waist", where there is a square cross-section. Each half is hollowed out. The halves are held together by magnets, of alternating polarity so one can only make the correct shape. In fact, the spacing between centers is rather more than $r\sqrt{2}$ as one does not want it to roll off the table! It is handsomely made in shiny metal and called Doublette. Ucke has kindly sent us an example.

Ucke goes on to discuss the Oloid and some variations, then looks at the paths and touching lines on the plane. The case of the twisted two half circles (i.e., the distance between centers is zero) can be calculated easily, but questions for the Wobbler remain unsolved.

Dirnböck and Stachel [9] find equations for the surface of the convex hull of the two circles, i.e., for the surface of the Oloid, hence they can explicitly express the rolling on a plane and they give equations for the curve of contact — computer algebra systems failed to do the relevant integrations! They also find the curve traversed by the center of gravity. The motion of Schatz's Turbula linkage is inverse to the movement with respect to the frame attached to the Oloid. When the defining circles have unit radius, the surface area of the Oloid is that of the unit sphere, i.e., 4π. They compute the volume numerically as 3.05241.... The authors say Schatz patented the Oloid in 1933, but this cites his first patent on linkages related to rotating rings of tetrahedra and has nothing to do with the Oloid, which was patented in 1970 [1]. In [10], the authors study the general problem of the movement of a rolling body and then do the Oloid as an example, citing [9] and repeating the incorrect citation to the 1933 patent.

Bibliography

[1] P. Schatz. Swiss patent 500,000: "Hilfsmittel zur Erzeugung einer taumelnden Bewegung." Applied: 3 August 1968; Issued: 15 December 1970; Published: 21 December 1970. 2pp + 1p diagrams. Also patented in France (FR2015012), Germany (DE1936595) and the USA (US3610587). (This had not been seen by us when the above article was written. Cited in [2].)

[2] G. Müller, ed. "Phänomena — Eine Dokumentation zur Ausstellung über Phänomene und Rätsel der Umwelt an der Seepromenade Zürichhorn, 12 Mai–4 November 1984." *American Journal of Physics* (1984) 79.

[3] A. T. Stewart. "Two-Circle Roller." *American Journal Physics* 34(2) (February 1966) 166–167.

[4] C. Ucke and H.-J. Schlichting. "Wobbler, Torkler oder Zwei-Scheiben-Roller." *Physik in unserer Zeit* 25(3) (1994) 127–128.

[5] C. Engelhardt and C. Ucke. "Zwei-Scheiben-Roller." Preprint of 3 May 1994, due to appearance in *Mathematisch-Naturwissenschaftlicher Unterricht*, 48(5) (1995) 259–263.

[6] C. Ucke. "Physics, Toys and Art." in *Quality Development in Teacher Education and Training*, M. Michelini, ed., 2004, 96–101.

[7] P. Schatz. *Rhythmusforschung und Technik*, 2nd ed. Verlag Freies Geistesleben, 1975.

[8] G. D. Hiscox. "Item 355: The pantanemone." *Mechanical Appliances Mechanical Movements and Novelties of Construction. A second volume to accompany his previous Mechanical Movements, Powers and Devices*, Norman W. Henley Publishing, 1904, 144.

[9]   H. Dirnböck and H. Stachel. "The Development of the Oloid." *Journal for Geometry and Graphics*, 1(2) (1997) 105–118.

[10] T. Warken and E. Schömer. "Rolling Rigid Objects." in *Intl Conf. in Central Europe on Computer Graphics, Visualization and Computer Vision, WSCG*, vol. 1, 2001, 57–62.

Chapter 12

Calculating for Fun

Getting students to carry out calculations can be difficult unless one embeds the calculations in an amusing or interesting setting. This chapter describes several calculations which have intrigued students for centuries and ought to intrigue students for centuries to come. This leads to large numbers and the interest of the problem is to make these numbers comprehensible. The problems include: the chessboard problem; the landowner's earth and air problem; the value of Manhattan Island; heavy rainfall; permutations of the alphabet and the number of crosswords; grains of sand versus stars in the sky; winning the lottery versus dying; the size of a million pounds. Each of these leads to large numbers and the interest of the problem is to make these numbers comprehensible. In addition, the problems lead to questions of measurement, area, volume, the size and population of the earth, interest rates, etc., and some research may be needed to determine the relevant numbers. Each student or group can be encouraged to produce improved measurements, variations on the method or variations on the problem itself. As a further variation, one can convert into other systems of units — in first year physics, we had to convert the speed of light into furlongs per fortnight.

12.1 The Chessboard Reward

Most readers and many students know some form of the legend[1] that the inventor of chess, when asked to name a reward by his sovereign,

---

[1]This section appeared in *Infinity*, 1 (April 2005) 14–16.

requested just enough rice to cover the chessboard — one grain on the first square, two on the second, four on the third, eight on the fourth, etc. The King was amazed by the modesty of the request and ordered it to be done forthwith. The next day, he was rather surprised to see a stream of porters carrying sacks of rice and rather distraught to hear they had only reached the 30th square. Sending for the Royal Mathematician, he learned that his largesse comprised

$$1 + 2 + 4 + 8 + \cdots + 2^{63} = 2^{64} - 1 = 18446744073709551615$$

$$\simeq 1.8 \times 10^{19} \text{ grains of rice.}$$

Many versions of the story then convert this gigantic number into something a bit more comprehensible. The first essential step is determining the size or weight of a grain of rice and this could be the first exercise to set to a class.

Using the basmati rice in our kitchen, we find that a rice grain is approximately cylindrical with an average diameter of 1.8 mm and average length 7 mm. This gives a volume of 18 mm^3. Since rice has a density roughly the same as water, one grain should weigh about 18 mg. To confirm this, we counted out 500 grains and weighed them, getting about 9 g. These measurements are pretty crude, though they are probably accurate to within 20% and this would be sufficient for our present purposes. The 500 grains were found to weigh 7.9 g, giving a weight of about 16 mg per grain. This leads to a relative density of 0.9, which is clearly an underestimate as rice sinks in water. Both the counting and the measurements could be improved — not all grains are the same size but we chose largish ones to measure, and there are broken grains which are difficult to combine into full grains. Both effects led to overestimating the volume of the bunch of 500 grains and hence underestimating the density.

We then tried to measure the density of rice directly, a la Archimedes. We weighed out 200 g and put it into a measuring cup with water in it. The water level rose 150 ml, giving a relative density of 1.3. This is rather larger than indicated by the previous result but further reflection leads us to think the difference has to do with the nature of rice or other grains. Basically, the grains are somewhat porous. Hence the volume determined by external measurement is somewhat larger than the volume of actual material enclosed. Imagine trying to determine the volume of tissue in a sponge by measuring its external dimensions! So the density determined by the previous method is very likely to be an underestimate even if the measuring and counting are accurate. This leads us to two values of the density — one based on actual volume of material and one based on the volume occupied. Students could carefully measure more grains to get a better estimate of the size and carefully count out more grains for a more precise measurement of the weight — perhaps finding different values for different kinds of rice, or for other kinds of grain. Then they could try measuring the density more accurately by the displacement method.

Before applying these data, we should note that cylindrical grains cannot pack tightly together. The best possible packing of cylinders seems most likely to be that with all the cylinders parallel and the circles in hexagonal packing. In this packing, the proportion of space occupied by the rice is the ratio of the area of a circle to the area of its circumscribed hexagon. Since both figures have the same inradius, r, the ratio of areas is the same as the ratio of their perimeters. If we let the side (= circumradius) of the hexagon be 1, then its perimeter is 6, the inradius $r = \frac{\sqrt{3}}{2}$ and the perimeter of the circle is $\sqrt{3}\pi$. Hence the optimal packing ratio is $\frac{\sqrt{3}}{6} = 0.91$. A measurement confirms this. We measured out 500 ml of rice and weighed it, getting 400 g. If the relative density of rice, with respect to the volume occupied, is taken as 1, the measurement indicates an actual packing ratio of 0.8, somewhat less than the theoretical optimum, as would be expected. The value of 1 is the more relevant relative density for rice since measuring out rice uses the occupied volume. The last measurement shows that the looseness of the packing reduces the relative density to the effective value of 0.8.

So we will not be too far off if we say that a grain of rice weighs 16 mg and occupies 20 mm^3 when in a pile. However, students might like to use an more accurate value of the density to find a more accurate actual packing ratio. (Joseph Vinot [20] estimates 20,000 grains of wheat per liter, which is 50 mm^3 per grain. Can other values be found?)

Thus the rice reward on a chessboard will have a volume of about 3.7×10^{20} mm$^3 = 3.7 \times 10^{11}$ m$^3 = 370$ km^3. This would make a cube about 7.2 km on an edge. One can perhaps visualize a cubic kilometer as the size of a small steep conical mountain, 1 km high, with sides at an angle of 45°. (This is estimating π as 3.) So 370 km^3 would be the same shape of mountain, but about 7 km high. (This is about $4\frac{1}{3}$ mi high.) Perhaps a group of students could try estimating the volume of a local mountain.

Early writers often compared the rice to the earth saying it would cover the earth's surface to some depth. Taking the earth's radius as 6400 km, the earth's surface area is 5.1×10^8 km^2 = 5.1×10^{14} m^2. Dividing this into the volume of rice gives an average depth of 0.7×10^{-3} $m = 0.7$ mm, which is barely noticeable. However, only about 30% of the earth's surface is land, so we can divide the previous result by 0.3 to get 2.3 mm depth if the rice is spread over just the land area. If we recall that the legendary origin of chess takes place in India, we could ask for the depth of rice when spread over India. My atlas says the area of India is about 3.3×10^6 km^2 = 3.3×10^{12} m^2, so the depth is 0.12 m or nearly 5 inches. Early writers often converted the rice reward into shiploads or fleets of ships, camel caravans, etc. Students may like to carry out some such conversion by determining the size of current grain ships, etc. How far would all the grains stretch? One might make use of the following.

Taking a density of 0.8 kg/l = 0.8 (metric) tons/m^3, the rice reward would have a mass of 3.0×10^{11} tons. An encyclopedia states that rice yields of 6 tons/ha were very good in the 1960s, so this can be taken as a rough estimate of current average yield. (Another source says world yields in 1948–1953 were 1.6 tons/ha.) The earth has about 1.5×10^{14} m^2 = 1.5×10^{10} ha of land area. If all of this were used to produce rice at the above yield, we would get 9×10^{10} tons per year, so it would take about $3\frac{1}{3}$ years for the entire world to produce enough rice. (Here and later, all tons are metric.) The encyclopedia also states that the world production of grain in the 1960s was about 1×10^9 tons/year, so the rice reward represents about 300 years of the whole world's grain production at that rate.

Viewed another way, the world's population is currently about 6×10^9, so the rice reward works out to about 0.62×10^2 = 62 m^3 of rice per person. This is the volume of a moderate-sized room. (Students can compute the volume of their classroom.) Taking the relative density of 0.8, this would have a mass of about 50 tons = 50,000 kg, a bit more than the capacity of the biggest lorries on our roads. Allowing a substantial 1 kg of rice as a daily diet, 50 tons of rice would feed one person for 50,000 days = 137 years, or two persons for their lifetime. Thus the total rice reward would feed twice the world's population for their entire life! Of course, this would be a boring diet and one would die of beri-beri or one of the numerous other diseases of malnutrition.

Inquiries have been made of colleagues in nutrition about the estimate of 1 kg per person day. Brown rice yields about 3.6 Calories/g and other grains have similar values. (Note that a Calorie is a kilocalorie.) A basic diet is 2000 to 3000 Calories/day. 2500 Calories per day requires about 700 g of rice per day. This value can be confirmed from the above-mentioned annual grain production in the 1960s being 1×10^9 tons. At the time, there were about 3×10^9 people, so the annual consumption was about $\frac{1}{3}$ ton per person or 900 g per day. However, a large portion of grain production goes to feed cattle! Students should check whether 6×10^9 people and 700 g of rice per day are still reasonable estimates.

Don Lemon [10] says two pounds of rice per day is the standard allowance to a laborer, where 2 lbs/day = 908 g/day. However, on p. 64, he says a Malay laborer gets 56 lbs per month and a Burmese or Siamese gets 46 lbs per month. Taking 365/12 days in a month, these amounts are 1.84 and 1.51 lb/day = 828 and 687 g/day. A letter in *Notes & Queries* [14] says subsistence is $\frac{1}{5}$ ton of grain per year, which converts to about 500 or 550 grams per day.

12.2   The Landowner's Earth and Air

Simple Simon has just bought a hectare (= $10^4 \, \mathrm{m}^2$) of land to build on. He is musing about the title deed which refers to mineral and air rights and he asks: "How much earth and how much air do I own?" Does he own more or less than the amount of the rice reward?

The earth problem is simpler. In theory Simon owns a conical shape, with vertex at the center of the earth and base having area $A = 1$ ha (hectare). (Actually, the area is measured on a spherical surface, but the curvature is not noticeable for such a small area and the difference doesn't affect the resulting volume.) Taking the radius of the earth as R = 6400 km, as before, and using 1 ha = $10^4 \, \mathrm{m}^2 = 10^{-2} \, \mathrm{km}^2$, he owns a volume V given by $V = AR/3 = 21\,\text{km}^3$. This really isn't so big — just the size of a 2.7 km high mountain. (While it is surprising how small this actually is, a hectare isn't all that big.) If Simon were to excavate his entire earthly possession, the resulting hole would only hold about 5.7% of the rice reward!

Now how much does Simon's earth weigh? This leads to the distinction between mass and weight which may mystify younger students, so perhaps one should rephrase it as: How much mass does Simon own? Here one needs to consult a reference book to find that the average relative density of the earth is $\rho = 5.522$, which we take as 5.5. That is, the mass of Simon's piece of the earth is 5.5 times the mass of the same volume of water. We know water has a mass of 1 kg per liter and a liter is $10^3\,\text{cm}^3 = 10^{-3}\,\text{m}^3$, so $1\,\text{m}^3$ of water has mass $1000\,\text{kg} = 1$ ton (metric) (actually this conversion has already been used above). So $V = 21\,\text{km}^3 = 21 \times 10^9\,\text{m}^3$ would contain a mass of 21×10^9 tons of water or 1.2×10^{11} tons of earth.

Alternatively, one can look up and find that the earth has mass $6.0 \times 10^{24}\,\text{kg}$ and surface area $5.1 \times 10^8\,\text{km}^2 = 5.1 \times 10^{10}$ ha. Thus the mass per hectare of surface is $1.2 \times 10^{14}\,\text{kg} = 1.2 \times 10^{11}$ tons, as before.

For the air, we again need to consult a reference book and mine gives an estimated mass of the earth's atmosphere as $5.2 \times 10^{18}\,\text{kg}$. Dividing this by the earth's surface area gives a mass of $1.0 \times 10^8\,\text{kg}$ of atmosphere for each hectare of land. (If the earth's atmosphere was homogeneous with the same density as at sea level, it would reach just up to 8.0 km.)

Again, one can obtain this result a quite different way. Consulting a reference once more, we see that a pressure of one atmosphere is equivalent to the weight of a mass of 1 kg on each square centimeter, i.e., $1\,\text{kg/cm}^2 = 1 \times 10^4\,\text{kg/m}^2 = 1 \times 10^8\,\text{kg/ha}$. (Incidentally, this tells us that each square meter has a mass of 10 tons of air pressing down on it. The human body has about $4\,\text{m}^2$ of surface area, so it is being pressed by some 40 tons of air! A crude estimation gives the value of $4\,\text{m}^2$ — students could try to find a better estimate. Other sources say the area is about $20\,\text{ft}^2 \simeq 1.86\,\text{m}^2$ or $3000\,\text{in}^2 = 20.83\ldots\,\text{ft}^2 \simeq 1.93\,\text{m}^2$.)

Of course, if we ask for the volume over Simon's land, then that is infinite (assuming the universe is actually infinite — if not, the space over Simon's land is not clearly defined).

Since we have already found, or can easily find, values for the earth's volume, mass, area and land area, and for the mass of the atmosphere, it is easy to determine how much is available for each of the 6×10^9 people on the earth.

Volume of the earth $= 1.1 \times 10^{12} \, \text{km}^3$.

Volume per person $= 180 \, \text{km}^3$.

Mass of the earth $= 6.0 \times 10^{24} \, \text{kg}$.

Mass per person $= 1 \times 10^{15} \, \text{kg}$.

Area of the earth $= 5.1 \times 10^8 \, \text{km}^2$.

Area per person $= 0.09 \, \text{km}^2 = 9 \, \text{ha}$.

Land area of the earth $= 1.5 \times 10^8 \, \text{km}^2$.

Land area per person $= 0.025 \, \text{km}^2 = 2.5 \, \text{ha}$ (not really very much).

Mass of the atmosphere $= 5.2 \times 10^{18} \, \text{kg}$.

Atmosphere per person $= 8.7 \times 10^8 \, \text{kg}$.

Volume of the seas $= 1.4 \times 10^3 \, \text{km}^3$.

Sea per person $= 230 \, \text{m}^3$.

The Isle of Wight has an area of $381 \, \text{km}^2 = 147.1 \, \text{mi}^2$. Could all the people in the world get onto it? Well, $381 \, \text{km}^2 = 3.81 \times 10^8 \, \text{m}^2$, so putting 6×10^9 people on it gives about $4/60 = 1/15 \, \text{m}^2$ per person. This is an area about 26 cm square and most people would barely fit onto it. As an exercise try packing a bunch of people into as small an area as possible; try to measure the area needed for a person; find a more appropriate island.

An 1891 joke book [11, p. 258] presents the case of a barber who had been shaving the beard of a largish gentleman for 21 years without payment and finally submitted a bill for 1d per day for 7670 days, making 31£ 9s 2d. The gentleman thought this was a bit steep, so the barber offered to charge by the acre, at £200 per acre. They made some measurements and agreed that the shaved area was 192 in^2, repeated 7670 times, making 1,472,640 in^2, coming to 46£ 19s 1d, which is 15£ 9s 11d more than the initial bill.

First, we observe that $7670 = 365 \times 21 + 5$, so there were five leap-years in the 21 years. This restricts the date on which the period must have started — we get that it could have started from first of March of a leap-year through 28th of February of the year before the next leap-year.

Second, $7670\text{d} = 639s\ 2\text{d} = 31£\ 19\text{s}\ 2\text{d}$, so we see there is an error or misprint in the initial bill. The incorrect amount was used in calculating the difference of the amounts, so it is not just a misprint, but must be an error in the calculations.

The area is correct, since $1{,}472{,}640\ \text{in}^2 = 10{,}226\frac{2}{3}\ \text{ft}^2 = 1136\frac{2}{27}\ \text{yd}^2$. There are $4840\ \text{yd}^2$ in an acre, so the area is 0.2348 acre. At £200 per acre, this costs $46.9544\ £ = 46£\ 19\text{s}\ 1.054\text{d}$. Rounding to the nearest pence agrees with the barber's second bill, and this is $14£\ 19\text{s}\ 11\text{d}$ more than the correct amount of the initial bill.

However, $192\ \text{in}^2$ is considerably more than a square foot and this seems rather large. Google turns up a site which says men shave $48\ \text{in}^2$ on average.

12.3　Buying Manhattan

In 1626, Peter Minuit purchased the Island of Manhattan for the Dutch West India Company from Chief Manhasset of the Canarsees (Indians based in Brooklyn) for axes, trinkets and cloth worth 60 guilders (about \$24). It is often said that the Indians were swindled. In fact, it appears that the Indians who sold the island weren't from Manhattan and had no rights to the island, so it was the Dutch who were swindled! A version of the story says the Dutch had to buy the island again some years later from the rightful owners [9]. But assuming the deal was legitimate, were the Indians swindled? What if they had invested their money?

If the Indians could have deposited their money in a savings account, or otherwise made a good investment, and let the interest compound at an annual rate of r, then the account would be worth $P_n = 24(1 + r)^n$ after n years. Assuming they did this at the end of 1626, so their first interest payment was at the end of 1627, then $n = 378$ years will have elapsed at the end of 2004. Unfortunately, there is no simple estimate of long-term average interest rates. But typical rates are in the range from 3% to 10%. Below P_{378}

is tabulated for each integer value of percentage interest rate in this range. We use M for million (10^6), B for billion in the sense of 10^9, T for trillion (10^{12}) and Q for quadrillion (10^{15}).

r	P_{378}		
3%	1,708,764 $		
4%	65,889,088 $	or	66 M$
5%	2,453,388,302 $	or	2.5 B$
6%	8.83×10^{10} $	or	88 B$
7%	3.07×10^{12} $	or	3.1 T$
8%	1.03×10^{14} $	or	100 T$
9%	3.37×10^{15} $	or	3.4 Q$
10%	1.06×10^{17} $	or	110 Q$

So we see the Indians would have a lot of money in their account. It has occasionally been asserted that they could buy back Manhattan, though this may just mean the land, not the buildings. The area of the island is about $23 \, \text{mi}^2 \simeq 1.5 \times 10^4$ acres. (Another source says 22 mi^2). We have not been able to find any definite estimate of the value of Manhattan as a whole. Reading through 1979 & 1994 Michelin Guides to New York, I finds two items of information which can be used to produce rather crude estimates.

In 1946, John D. Rockefeller Jr. gave 8.5 M$ to buy the 16 acre site for the UN, giving a value of about 0.5 M$/acre. At the time, the site was basically slums, but much of Manhattan was slums and considerably further from the center than the UN site, so this may be a reasonable average value for the island as a whole, giving a total value of 7 B$ in 1946. In 1946, after 320 years, an investment of $24 at 6% compound interest would be worth 3 B$, while if 7% were obtained, the investment would be worth 61 B$.

About 1979, the 22 acres of Rockefeller Center produced a land rental income of 18 M$ per year. Enquiry to the estate agents Knight Frank revealed that the value of land varies roughly between 8 to 16 times its annual rental, with the larger value applying in more favorable areas where the land is in demand and easy to dispose of. So the value of the land of Rockefeller Center was about 288 M$, giving a value of about 13 M$/acre. However, this is among the most expensive land in Manhattan, so an average value might be in the order of 1 M\$/acre, giving a total value for the island of about 15 B\$. Since 1979 is 343 years after 1636, we compute the value of \$24 after 343 years at 6% and get 11 B\$, while the value at 7% would be 288 B\$.

Knight Frank also said that a cleared office site in the middle of the City of London sold in late 1995 for $3\frac{1}{4}$ M£ for an area of about 3000 ft^2. An acre is 43,560 ft^2, which gives a value of 47 M£/acre. At the then current exchange rate of 1.5\$/£, this converts to about 71 M\$/acre. Allowing for inflation, this indicates that the above estimates are of the right order of magnitude.

So we see that the Indians' investment would certainly have given them approximately enough money to buy back the land of Manhattan, provided they got at least 6% return.

However, the buildings of Manhattan are worth many times the value of the land — a 1986 report that a typical office block in central Manhattan was worth 80 M\$. But if the Indians had obtained 7% return, they could buy back a fair amount of the buildings as well as the land.

The famous Canadian humorist and economist, Stephen Leacock, discusses the Manhattan Island question [8]. His details are a bit different than history records, but the idea is the same. He says the Dutch bought Manhattan for \$50 in 1621, but he states that interest rates were quite high at the time: "people got 10 per cent easy enough for business loans (kings paid up to 20 or more), and even a hundred years later 8 per cent. was easy enough." He consequently assumes 10% up to 1735, then 8% until 1835, 6% until 1895, 5% to 1923 and 4% until 1943. He obtains an overall factor of 235 yielding "something more than one and a half trillion dollars", i.e., 1.5 T\$ and asserts that this "is 240 times the assessed value of all the land and all the property in Manhattan. It would buy all the United States." We get the following $1.1^{114} \cdot 1.08^{100} \cdot 1.06^{60} \cdot 1.05^{28} \cdot 1.04^{20} = 3.26183 \times 10^{10}$, which multiplied by 50\$ is 1.631 T\$, while $2^{35} = 3.43597 \times 10^{10}$, which multiplied by 50\$ is 1.718 T\$, either of which is "something more than" 1.5 T\$. Rather than using a detailed value of 2^{35}, he might well have used the common approximation $2^{10} \simeq 10^3$, getting $2^{35} \simeq 25 \times 10^9 \cong 3.2 \times 10^{10}$ which multiplied by 50\$ would give 1.6 T\$. From the multiplier 3.26183×10^{10}, we can determine the average interest rate as 7.81%. Continuing his calculations to the end of 2004 at 4%, we get a multiplier of 35.68577×10^{10} and a value of 17.843 T\$. Leacock's data indicates the assessed value of Manhattan and its buildings was 6.25 B\$ in 1943, which seems a bit low compared to the previous estimates, but assessed values are often only a fraction of the real value.

Leacock says further that at 10%, money doubles in seven and a quarter years, etc., and his double-lives are reasonable approximations. This leads to the question of estimating the doubling period. If our interest rate is r, then the double-life, d, is the solution of $(1 + r)^d = 2$, i.e., $d = \log 2/\log(1 + r)$, where any logarithm can be used. However, this is a difficult calculation without a calculator, so accountants and others evolved the following approximation. The basic approximation is that $\ln(1 + r) \simeq r$. (This can be established in many ways, but these use some aspect of the calculus and hence are beyond the present discussion. However, if you know that $e^r = 1 + r + r^2/2! + r^3/3! + \cdots$, it is clear that for small r, $e^r \simeq 1 + r$ and taking natural logarithms gives us $r \simeq \ln(1 + r)$.) Applying this

$r\%$	exact d	$69.3/r\%$	$70/r\%$	Leacock
1	69.66	69.3	70	
2	35.00	34.65	35	
3	23.45	23.1	23.33	
4	17.67	17.33	17.5	$17\frac{1}{2}$
5	14.21	13.86	14	14
6	11.90	11.55	11.67	12
7	10.24	9.9	10	
8	9.01	8.66	8.75	9
9	8.04	7.7	7.78	
10	7.27	6.93	7	$7\frac{1}{4}$
11	6.64	6.3	6.36	
12	6.12	5.78	5.83	
13	5.67	5.33	5.38	
14	5.29	4.95	5	
15	4.96	4.62	4.67	
16	4.67	4.33	4.38	
17	4.41	4.08	4.12	
18	4.19	3.85	3.89	
19	3.98	3.65	3.68	
20	3.80	3.47	3.5	

to our problem gives us $d \simeq \ln 2/r$ and $\ln 2 = 0.693\ldots \simeq 0.7$, so $d \simeq 0.7/r$. If we express r as a percentage, say $r\% = 100 \cdot r$, then $d \simeq 70/r\%$. Above is a table of the exact and approximate double-lives for $r\% = 1, 2, \ldots, 20$. Also given are the estimates stated by Leacock for the interest rates he mentions.

From the table we see that $69.3/r\%$ is an underestimate, corresponding to the fact that e^r is an overestimate of $1 + r$. Taking $70/r\%$ improves the estimate slightly, but is still an underestimate except at 1% and 2% and is accurate to the nearest year except at 11%, 13%, 16%. To compensate for this error, accountants have increased the numerator a bit and taken $d \simeq 72/r\%$, which is almost exact at 8% and an overestimate for lower rates and an underestimate for higher rates. This has the advantages that 72 has lots of factors and that the estimate is quite accurate for the range of interest rates in current use.

Returning to Leacock — how did he find his estimates? He did not have the advantage of electronic calculators, so perhaps he used some formula like $d \simeq 72/r\%$, rounding the results. However, his estimates are not consistent with this formula for numerators 70, 71, 72. Although he presumably did not calculate the exact values of d for such a light piece of writing, tables of these values had been available for some centuries and he would have been able to look up the exact values of d and round them. However, this still doesn't seem right, as he has given some values to the nearest quarter and rounding of the values for 4% and 5% would have given him $17\frac{3}{4}$ and $14\frac{1}{4}$. Perhaps he was using some rule of thumb like taking $70/r\%$ for rates up to 5% and $72/r\%$ for higher rates.

Again assuming that Leacock did not carry out the detailed calculations, how did he come to his figure of 35 doublings? The story would seem to indicate that he considered 10% interest for 114 years, and since money doubles every $7\frac{1}{4}$ years at 10%, then there would be $114/7.25 = 15.72$ doublings in this period. Using this method, there would be a total of $114/7.25 + 100/9 + 60/12 + 28/14 + 20/17.5 = 34.978$ doublings in this period, so his assertion of 35 doublings is pretty reasonable.

Here are two other statements about early interest rates. A London guide [16] says that before the establishment of the Bank of England in 1694, the government paid 10 to 12 per cent while private borrowers paid 20 to 30 per cent. Another guide [6], discussing the foundation of the Bank of England, states that kings and their ministers no longer had to beg money in the City, paying interest of 15 per cent upwards. The Bank loaned its capital to the government at 8 per cent, but with time this reduced to 3 per cent on the Government Consolidated Stock — the "Consols" mentioned throughout the 19th century. On 5 April 1892, the rate was reduced to $2\frac{3}{4}$ per cent.

There are only a few other cases where territory has actually been bought by a country. It is not easy to place a definite value on the land at present but students may like to work out the cost per square mile or acre for these purchases and the current value of the cost price assuming 6% or 7% interest.

- In the 18th century, the Viscount of Turenne sold his viscounty to Louis XV for 4,200,000 livres (about £1,500,000 according to a 1965 guide-book), but the area involved is unknown [12].
- In 1741, the inhabitants of the Danish island of Fan purchased the island from the crown at a public auction for 8200 Rigsdaler. The island is $10\frac{1}{4}$ miles long and had 2575 inhabitants in the early 1950s [1].
- The Isle of Man long had a separate sovereign. In 1765, the British government purchased the sovereignty from the Duke of Atholl for £70,000 plus an annual payment of £2,000 per year. The Isle has an area of 227 mi^2. However, the Atholls retained manorial rights, including the patronage of the see. These were purchased for £417,144 in 1828. Details have come from the *Encyclopedia Britannica*, which does not indicate when, if ever, the annual payment ceased. Assuming it started in 1766 and ceased in 1828, there were 63 payments, totalling £126,000. A crude estimate would be that the cost was £70,000 + 126,000 + 417,144 = £613,144. However, the payments were made at different times, so one ought to convert all of them to 1828 values. Interest rates at the time seem to have been about 3%. This gives $70000 \cdot 1.03^{63} + 2000 \cdot \frac{(1.03^{63}-1)}{0.03} + 417144 =$ £1,230,324.97 as the 1828 value of all the payments.
- In 1803, the US made the Louisiana Purchase of 828,000 mi^2 from financially pressed Napoleonic France for 27 M\$. This purchase comprises the middle third of the US. Most of it is farm land, but it does include Chicago, St. Louis and New Orleans.

- In c. 1835, Hardy Ivy bought $202\frac{1}{2}$ acres for $225, possibly from the local Indians. This comprises downtown Atlanta [3].
- In 1854, the US made the Gadsden Purchase from Mexico. The treaty ending the Mexican War in 1848 set a boundary according to a crude map and the Purchase rectified the boundary, obtaining the territory due west of El Paso for the US and a possible railroad. The area is about a quarter of the state of Arizona, comprising 29,640 mi^2 and cost 10 M$.
- On 30 Mar 1867, the US bought Alaska from the Russians for 7.2 M$. The area is 586,000 mi^2.
- In 1898, the US bought the Philippines from Spain for 20 M$. They spread over some 300K km^2 (= 116K mi^2), which seems to count just the land area.
- In 1917, the US bought the Danish Virgin Islands from Denmark for 25 M$. The total area is only 266 mi^2, but they had great strategic importance.
- Parts of Monaco have been bought and sold, but the price is unclear. The adjacent town of Menton was purchased by the Grimaldis of Monaco in the mid-14th century. In 1860, Menton and Rocquebrune were purchased from Monaco by France.
- In 1649, John, first Lord Culpepper, was granted more than 5M acres of Virginia by Charles II, in return for an annual rental of 6£ 13s 4d = 6.67£ [5].
- In 1805, Johannes Coelombie, of Haarlem, Netherlands, left a legacy of 16,000 guilders to accumulate until 140 years after the death of his 21-year old servant. This occurred on 4 Jan 1999, and the value of the legacy was 9,200,000 guilders (about 3 M£). This corresponds to a multiplication of exactly 575 in 193 years which corresponds to an interest rate of 3.347% for the period [7].

These are the only examples of long term interest of this sort, known to us.

12.4 "It's a Hard Rain a Gonna Fall!"

We all know that a heavy rainfall can produce an inch or more of rain in a short time. But how much rain is this? How much actually lands on our backyard or on our roof?

People have quite different sizes of backyard and house, so we will take a moderate sized imaginary yard $10\,\text{m} \times 10\,\text{m}$, giving an area of $100\,\text{m}^2$. For a moderately heavy rainfall of $1\,\text{cm}$, the volume of rain is $1\,\text{m}^3$ and recall that this weighs a ton! Thank God it all doesn't come down at once!

A 1955 quiz book contains: How much does one inch of rain on an acre weigh? It's not hard to convert this to the previous form if we know that $1\,\text{in} = 2.54\,\text{cm}$, $1\,\text{acre} = 4840\,\text{yd}^2 = 4046.87\,\text{m}^2$, so the volume of rain is $2.54 \cdot 40.4687\,\text{m}^3 = 102.79\,\text{m}^3$, which weighs 102.7 tons.

One of the most peculiar units of measurement is the acre-foot, which is a measure of irrigation water in the American west and hence a matter of great economic interest. It is twelve times the quantity just computed, i.e., $1233.5\,\text{m}^3$ or 1233 tons.

A recent book asserted that Robespierre's fate was sealed when "a million tons of water" fell on Paris. A letter then asserted that someone had checked the weather reports and found no record of rain on the day. But is a million tons a reasonable amount or just an author's guess? An encyclopedia gives the area of Paris as about $105\,\text{km}^2$ in about 1970. Paris is contained within the $78\,\text{km}^2$ area inside the *boulevard périphérique*, which seems to be ignoring the outer suburbs [15]. The area at the time of the French Revolution would have been much less, let's say about 20 km². On this area $1\,\text{cm}$ of rain would be $0.2 \times 10^6\,\text{m}^3$ or a fifth of a million tons of water. So a million tons would require $5\,\text{cm}$ of rain, which is a substantial rainfall. Even assuming an area of $50\,\text{km}^2$ requires $2\,\text{cm}$ of rain, which is a fair amount. So we see that a million tons of water is a feasible amount and it could not have been unobserved.

12.5 Permutations and the Number of Crosswords

Several 19th century puzzle books give the number of different ways of writing the 26 letters of the alphabet and how much space it would take to write or print all of these. The number of ways is simply given by

$$26! = 1 \cdot 2 \cdot 3 \cdot \ldots \cdot 26 = 403\,291\,461\,126\,605\,635\,584\,000\,000$$
$$= 4.03291 \times 10^{26}.$$

Each of these permutations contains 26 letters, so there are

$$26 \cdot 26! = 10\,485\,577\,989\,291\,746\,525\,184\,000\,000 = 1.0486 \times 10^{28}$$

characters to be written.

Since this is the modern age, we will not write these by hand, but use a computer. Using somewhat smaller than usual letters, we can print 8 lines to the inch and 15 characters to the inch, so each character occupies $1/120 \text{ in}^2$ and we will need

$$87\,379\,816\,577\,431\,221\,043\,200\,000 = 8.7380 \times 10^{25} \text{in}^2$$

of paper to print on. This is

$$606\,804\,281\,787\,716\,812\,800\,000 = 6.0680 \times 10^{23} \text{ ft}^2$$
$$= 67\,422\,697\,976\,412\,979\,200\,000 = 6.7423 \times 10^{22} \text{ yd}^2$$
$$= 21\,766\,108\,592\,592\,000 = 2.1766 \times 10^{16} \text{ mi}^2.$$

Using $2.54\,\text{cm} = 1$ in, we have $6.4516\,\text{cm}^2 = 1 \text{ in}^2$, and the required area is

$$56\,373\,962\,463\,095\,526 = 5.6374 \times 10^{16} \text{ km}^2.$$

Taking 4000 mi or 6400 km for the earth's radius, we get that the required area is about 109M times the earth's surface area! How much volume would this take? – after all, paper is pretty thin. With a calipers, we found that a ream (= 500 sheets) of standard A4 paper $(80\,\text{gm/m}^2)$ is 52 mm thick, so each sheet is 0.104 mm thick. Hence our area, $5.63740 \times 10^{16} \text{ km}^2$, of paper would have a volume of $5.86289 \times 10^9 \text{ km}^3$. 109M thicknesses of paper spread over the earth's surface would be just 11.336 km thick, only a bit higher than Mt. Everest. The paper would weigh 4.50992×10^{18} metric tons, which is only about $0.754 \times 10^{-3} = 0.000753 = 0.075\%$ of the earth's mass.

[The thickness of a folded piece of paper is often given as a question. For example, it is asserted that a sheet of newspaper folded 40 times would be 26,000 miles thick [2]. Folding a sheet 40 times gives a pile $2^{40} = 1.09951 \times 10^{12}$ sheets thick. At 0.104 mm per sheet, this would be $1.14349 \times 10^5 \text{ km} \simeq 71,053$ mi thick. But newspaper is somewhat thinner. Three measurements of newspaper give an average value of about 0.065 mm per sheet, so 2^{40} sheets are about

7.15×10^4 km $\simeq$ 44,408 mi thick, so Blundell's value [2] is not unreasonable. Taking his value, we find 26,000 mi $\simeq$ 41,842 km divided by 2^{40} gives about 0.038 mm per sheet.]

More interesting, we might ask how long the printing (or even just the calculating, since printing is much slower) would take. If we have a modern machine, we might be able to get it to compute 10^7 permutations per second. So we would need 4.03291×10^{19} sec, which is 1.27798×10^{12} years, which is about a hundred times the estimated age of the universe. However, the number of computers on earth is probably about 10^9, so if we could use all of these on our problem, we could get done in about a millennium!

A similar problem occurs in an Indian book [19]. How many ways are there to fill in a crossword puzzle? This has several answers, depending on how you treat the black squares. In the simplest case, we consider a typical British diagram, which is 15 × 15 and has about 160 white squares. The white squares can be filled with the 26 letters in $26^{160} \simeq 2.4873 \times 10^{226}$ ways. (It is most unlikely that your calculator can do this as most calculators permit at most two digits for the exponent. One needs to use logarithms and Sterling's approximation (see below and Appendix to Chap. 2).). This result is vastly greater than any other number yet seen in this chapter. The closest number is Eddington's estimate of the number of particles in the universe as 10^{79}. Uday [19] compared the number with the lifetime of the universe in seconds, but the universe is perhaps ten billion years old, i.e., 10^{10} years, and there are $31,536,000 = 3.1536 \times 10^7$ seconds in a year, so the universe has existed for 3.1536×10^{17} seconds, which is not very big at all in the present discussion.

Instead of filling in a given diagram, we can ask how many ways one can place the 26 letters and a black square into an $n \times n$ array. Then each square can have one of 27 values and the number of ways is 27^{n^2}. Computing this again requires logarithms. The values for a few values of n are given below.

$n =$	11	12	13	14	15
	1.57×10^{173}	1.31×10^{206}	7.95×10^{241}	3.53×10^{280}	1.14×10^{322}.

For the case $n = 15$, this increases the number of ways by almost 10^{100}! We can consider the problem on the 15 × 15 grid in another way. First we place the 65 black squares on the diagram and then we fill in the 160 white squares. This will be $\binom{225}{65}$ times our previous answer of 26^{160}. Again, this is difficult to calculate, not only because it is large, but because 225! would seem to require 224 multiplications.

Fortunately, there is one of the most useful approximations in mathematics, Stirling's approximation: $n! \simeq \sqrt{2\pi n} \cdot (\frac{n}{e})^n$, where e is the base of natural logarithms $e = 2.71828\,18284\,59045\,23536\ldots$. The relative error in Stirling's approximation decreases as n increases and is already less than 1% at $n = 9$. Using this and logarithms, we find

$$225! \simeq 1.26 \times 10^{433}; \quad 160! \simeq 4.71 \times 10^{284}; \quad 65! \simeq 8.24 \times 10^{90};$$

so our binomial coefficient $\binom{225}{65} \simeq 3.24 \times 10^{57}$. Multiplying this by 26^{160} gives 8.07×10^{283} ways to allocate 160 white cells on a 15×15 grid and then fill in the white cells.

One can reduce this number a bit because the pattern of black and white cells is normally symmetric about the center of the board, so we want to know how many ways there are to dispose about 33 black cells among the 113 cells of half a board, which leads to $\binom{113}{33} \simeq 1.19 \times 10^{29}$ and multiplying this by 2^{160} gives 2.97×10^{255} ways.

12.6 Grains of Sand versus Stars in the Sky

According to [18], Neil D. Tyson, an astrophysicist at Princeton, did some measurements and an estimate on sand grains. He went to Jones Beach on Long Island and found he could place 25 grains of sand along a centimeter, i.e., the diameter was 0.4 mm and the volume of a sphere of this size is $0.033\,\text{mm}^3 = 33 \times 10^{-12}\,\text{m}^3$. If these are arranged in a cubic lattice, so that each grain corresponds to a box of edge 0.04 cm, then there are 15,625 grains per cm^3 or 15.6×10^9 grains per m^3 or 15.6×10^{18} grains per km^3. Tyson then estimated a typical beach had volume $0.075\,\text{km}^3$ and hence had about 1.17×10^{18} grains of sand. Tyson estimated there are 10^{21} stars in the universe. Thus there are more stars in the sky than grains of sand on a beach, but it seems likely that there are over a thousand beaches and we can include all the deserts, so it seems clear there are more grains of sand on earth, or probably in the Sahara Desert, than stars in the sky. Another source says it is estimated that there are 10^{24} grains of sand on all the beaches of the world [13]. It has been asserted "there are as many cells in a brain as there are stars in the firmament, but it weighs only about three pounds (or for the European-minded, one-and-a-half kilograms)". [17]

The brain, like most tissues, is largely water. Water has molecular weight about 18, so 18 g of water is a mole and contains Avogadro's number of molecules, about 6×10^{23}. So 1.5 kg of water contains about 5×10^{25} molecules of water. If there are 10^{21} stars, Della Sala's assertion [17] gives us that a brain cell is about as heavy as $5 \times 10^4 = 50,000$ molecules of water. This seems like an underestimate as cells contain a lot of complex molecules, like DNA, with molecular weights in the tens and hundreds of thousands.

The brain is estimated to contain about 10^{10} (or 10^{11}) neurons. Dividing these into 1.5 kg, we find that a neuron weighs about 0.15 (or 0.015) μg. Since a molecule of water weighs $\frac{18}{6 \times 10^{23}} = 3 \times 10^{-23}$ g, a neuron is equivalent to about 0.5 (or 0.05) $\times 10^{16}$ molecules of water.

12.7 "A Lottery is a Tax on the Innumerate."

It has been said that one is more likely to die before the lottery is drawn than one is likely to win it! Let's see if this is true. To win the English lottery, one must correctly pick six numbers from 49 numbers. The number of combinations of 49 things taken six at a time is the binomial coefficient

$$\binom{49}{6} = \frac{49!}{6!43!} = \frac{49 \cdot 48 \cdot 47 \cdot 46 \cdot 45 \cdot 44}{720} = 13,983,816 \simeq 1.4 \times 10^7$$

and the probability of winning the lottery is the reciprocal of this value.

The current population of England is about 50M $\simeq 5 \times 10^7$. About $\frac{1}{2}$M $= 0.5 \times 10^6$ people die in England each year. This is about 9615 deaths per week $\simeq 10^4$. The probability of dying in a given week is thus $\frac{10^4}{5 \times 10^7} = \frac{1}{5000}$. (This assumes equal likelihood over all people, which is not reasonable when looking at a given person.) The ratio of this probability to the probability of winning the lottery is $\frac{13983816}{5000} \simeq 2797 \simeq 2800$. That is, you are 2800 times more likely to die in a given week than to win the lottery.

However, you might decide to improve your odds by not buying your ticket until late in the week. A week is 7 days = 168 hours = 10,080 minutes. Assuming the time of death is equally distributed, you would have to buy your ticket $10080/2797 = 3.60$ minutes = 3 min 36 sec before the draw in order to have your probability of dying before the draw equal to your probability of winning.

12.8   Storing a Million Pounds

In case you win on "Who Wants to be a Millionaire?", you may be interested to know how much that would be in real money, i.e., in actual 1£ coins. The current 1£ coin is $\frac{7}{8}$ in. (22.225 mm) in diameter, $\frac{1}{8}$ in. (3.175 mm) in thickness and $\frac{1}{3}$ oz (9.45 gm) in weight. Side by side, a million would stretch 22.2 km; if arranged in a square, that would have a thousand on each side and be 22.2 m on a side. Stacked in a single pile, they would be 3.2 km high. They would weigh 9.45 tons. If we stack them on a square base, we can form a roughly cubical shape with edge about 1.16 m. This would fit on an array of 52×53 coins and be 362 or, mostly, 363 coins tall.

Some additional exercises: What is the density of a 1£ coin? What if one is paid in 5£ notes, or 50£ notes, or even in 1p coins? How many coins are required to reach the moon? Do the same calculation in US money.

12.9   A4 Paper

A4 is part of one the most rational sizing schemes ever devised. Two A4 sheets make an A3 sheet and an A3 sheet has the same proportion as an A4 sheet — otherwise it would be a problem to reduce or enlarge between different sizes. See Figure 12.1.

Thus we want $L/W = W/(L/2)$, which gives us $L^2 = 2W^2$, or $L = W\sqrt{2} = 1.4142W$. This gives us the pattern of consecutive sizes where each size is twice the area and hence $\sqrt{2}$ times the dimensions of the next smaller size. The sequence is then completely specified by an appropriate arbitrary choice and this is done by specifying that A0 paper has an area of $1\,\text{m}^2$. Then the dimensions of A0 paper satisfy $LW = 1, L = \sqrt{2}W$, and these imply that $L = 2^{1/4} = 1.1892\,\text{m}$,

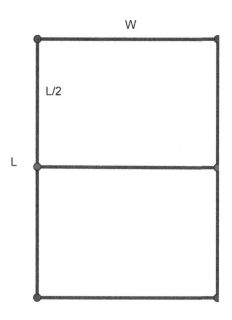

Figure 12.1.

$W = 2^{-1/4} = 0.84090 \, \text{m}$. Thus the dimensions of An paper are given as follows: This was first noted by Georg Christoph Lichtenberg in 1786,

A0 1189 mm × 841 mm Poster size

A1 841 mm × 595 mm Small poster size

A2 595 mm × 420 mm Approximately broadsheet newspaper
                    page size

A3 420 mm × 297 mm Approximately tabloid newspaper size

A4 297 mm × 210 mm Writing/typing paper size —
                    US writing paper is 279 × 216 mm

A5 210 mm × 149 mm Brochure size

A6 149 mm × 105 mm Postcard size

For metrophobics, the dimensions are given in inches and for decimophobics, they are rounded to the nearest eighth of an inch.

$$A0 \quad 46.81 \quad \times \quad 33.11 \quad 46\frac{3}{4} \quad \times \quad 31\frac{1}{8}$$

$$A1 \quad 33.11 \quad \times \quad 23.43 \quad 31\frac{1}{8} \quad \times \quad 23\frac{3}{8}$$

$$A2 \quad 23.43 \quad \times \quad 16.54 \quad 23\frac{3}{8} \quad \times \quad 16\frac{1}{2}$$

$$A3 \quad 16.54 \quad \times \quad 11.69 \quad 16\frac{1}{2} \quad \times \quad 11\frac{3}{4}$$

$$A4 \quad 11.69 \quad \times \quad 8.27 \quad 11\frac{3}{4} \quad \times \quad 8\frac{1}{4}$$

—US writing paper is $11 \times 8\frac{1}{2}$ inches

$$A5 \quad 8.27 \quad \times \quad 5.87 \quad 8\frac{1}{4} \quad \times \quad 5\frac{7}{8}$$

$$A6 \quad 5.87 \quad \times \quad 4.13 \quad 5\frac{1}{8} \quad \times \quad 4\frac{1}{8}$$

Since standard paper generally weighs $80\,\mathrm{g/m^2}$, it follows that an A0 sheet weighs 80 g, an A1 sheet weighs 40 g, ..., an A4 sheet weighs 5 g — a useful fact for those who mail letters.

For those who need peculiar sizes of paper, there is also a B series, where the area of B0 paper is $1.5\,\mathrm{m^2}$, so Bn paper has dimensions equal to $\sqrt{1.5} = 1.2247$ times the dimensions of An paper. We were told that this was wrong and that the area of B0 paper was $\sqrt{2}\,\mathrm{m^2}$, which would make the dimensions of Bn equal to $2^{1/4} = 1.1892$ times the dimensions of An. However, we have some B4 paper in its package and its dimensions are $257 \times 364\,\mathrm{mm}$ which gives $257/210 = 1.2238$, $364/297 = 1.2256$ and these seem clearly closer to $\sqrt{1.5}$ than $2^{1/4}$. Assuming this, then Bn paper dimensions would be constructed starting from B0 paper as follows:

$$B0 \quad 1456\,\mathrm{mm} \times 1030\,\mathrm{mm}$$

$$B1 \quad 1030\,\mathrm{mm} \times 728\,\mathrm{mm}$$

$$B2 \quad 728\,\mathrm{mm} \times 515\,\mathrm{mm}$$

$$B3 \quad 515\,\mathrm{mm} \times 364\,\mathrm{mm}$$

$$B4 \quad 364\,\mathrm{mm} \times 257\,\mathrm{mm}$$

This gives precisely the observed dimensions of B4 paper. If one assumes B0 paper has area $\sqrt{2}$, then its dimensions would be 1414×1000 mm and B4 paper would be 354×250 mm. The difference between these dimensions and the observed dimensions is small but certainly noticeable — 10 mm = 0.4 in.

There is a C series, but it is not clear how it would fit in, though one could fit in two series, with initial areas of 1.25 m^2 and 1.75 m^2. A printer told us that the C series is about 10% larger than the A series in order to permit cutting of edges after folding a large sheet into a signature, though it is unclear if the 10% refers to edges or area. There are D sizes for envelopes which have to be a bit bigger than the sheets that go into them.

In the 19th century, there were attempts to quantify aesthetics and it was claimed that a rectangle whose sides were in the "Golden Mean" was the most aesthetically pleasing. This ratio is $(1 + \sqrt{5})/2 = 1.618$, and many common papers sizes have ratios which are close to this — e.g.,

5 in by 3 in index cards have $L/W = 1.666$;

8 in by 5 in index cards have $L/W = 1.6$;

American legal paper, 14 in by $8\frac{1}{2}$ in, has $L/W = 1.647$;

American writing paper, 11 in by $8\frac{1}{2}$ in, has $L/W = 1.294$.

As we have seen, the ratio for A4 is $L/W = 1.414$, which is not too far from 1.618 and it certainly looks better than American writing paper.

12.10 Other Exercises

We have a newspaper clipping which asserts "You will spend three years of your life on the lavatory." Is this reasonable?

An average life span is now seventy-some years. If we assume 72 years, then 3 years is 1/24 of your life, i.e., an hour per day. This seems several times too large to me.

Jonathan Clements [4] asserts an average man spends 3500 hours shaving and removes about 30 ft of whiskers. What do these work out to per day? Men start shaving at about 15 and an average lifetime is about 75 years, so let us take a shaving lifetime of 60 years. Taking 365.25 days per year, 60 years is 21,915 days. 3500 hours is 210,000 minutes, so the time works out to $210,000/21915 = 9.58$ minutes per day, which seems reasonable. The growth is 360 in, so the growth per day is $360/21915 = 0.0164$ in, which is about $1/60$ inch per day, or 6 inches per year. Again, this seems reasonable. Another source also gives 6 inches per year for a beard and a friend who shaved off his beard two years ago and has let it grow says that agrees with his measurements.

However, Ref. [2, pp. 188 & 302] asserts a man grows about 94 miles of whiskers in a lifetime. This must be referring to the total growth, not just the length of beard. 94 mi = 496,320 ft, divided by 30 ft per whisker, gives 16,544 whiskers. Some estimates for the number of hairs on a human scalp vary from 40,000 to 100,000, while Google turns up several statements ranging from 90,000 to 150,000, so 16,544 may be reasonable for hairs in a beard. (Google also provides a statement on the number of hairs in a beard, which varies from 7,000 to 15,000.)

A friend said he recently saw the statement that you see several million images in your life. He said this was rather an underestimate. How many images does one see? You may know that film and TV refresh images 24 or 30 times per second so that you do not see any flicker. The time of persistence of vision is about 0.1 sec, so one can estimate that one perceives at most 10 images per second, and perhaps one per second would be better. There are about 3.16×10^7 sec in a year. Allowing about a third of the time for sleep, this still gives at least about 20M images per year and hence in a lifetime of 72 years, at least about 1.44B images in a lifetime.

Assuming there was a land route, could you walk around the earth? The equatorial circumference of the earth is 40076.594 km = 24902.400 mi. A vigorous hiker can cover up to 50 mi per day, but let's take 25 mi per day as a comfortable pace. This then requires about 1000 days $\simeq 2.72$ years, which is not impossible. Reference [2, pp. 72 & 244] says it would take one year, which assumes one can cover 68.2 miles per day.

In 2000 years, how many days are there? How many hours? minutes? seconds? When was a million days ago? Take the number of years since your favorite calendar started and compute the elapsed time. A selection of calendar starting dates is given below. Are there any others?

The Julian calendar starts at noon on 1 Jan 4713 BC. This is an artificial calendar created for astronomical purposes in 1582 and named after the creator's father, Julius Caesar Scaliger.

- The early Egyptian calendar may have started in 4214 BC.
- Archbishop Ussher computed the date of creation as about 18:00 on 22 Oct 4004 BC.
- The Jewish calendar starts with creation on 7 Oct 3761 BC.
- The oldest Hindu calendar, the Kali-Yuga, starts in 3101 BC.
- The ancient Greek calendar began with the first Olympiad in 776 BC.
- The ancient Roman calendar began with the founding of Rome in 752 BC, but this date was not universally accepted.
- The Buddhist calendar begins with the death of Buddha in 543 BC.
- The Jain calendar begins with the death of their founder Vardhamana in 527 BC.
- The Indian Vikrama calendar begins on 23 Feb 57 BC.
- The Indian Saka calendar begins on 3 Mar 78 AD.
- The Coptic calendar begins on 29 Aug 284 AD.
- The Armenian calendar begins on 9 Jul 552 AD.
- The Moslem calendar begins the day after Mohammed's Hegira, i.e., on 16 Jul 622 AD.
- The Parsee or Zoroastrian calendar begins on 16 Jun 632 AD.
- The French Revolutionary Calendar began on 22 Sep 1792 AD.

Modern computers have circuitry which works in nanosecond ($=10^{-9}$ sec) times. How far does light travel in a nanosecond? How does this affect the design of computers? (The speed of light is now defined as 299,792,458 m/sec = 186,282.3971... mi/sec. This determines the meter since the second is now defined as "9,192,631,770 periods of the radiation corresponding to the transition between the two hyperfine levels of the ground state of the caesium-133 atom.")

Many readers may know that $2^{64} - 1$ is also the number of moves required to move a Tower of Hanoi of 64 discs. If one can make one move per second, how long will it take? Suppose one can use a computer and make 10^9 moves per second, how long will this take?

Gold is remarkably ductile. Just $1\,\mathrm{mm}^3$ can be beaten into a square gold leaf $10\,\mathrm{cm}$ on an edge. How thin is it?

Oil spreads on the surface of water and has a smoothing effect. Benjamin Franklin made the first experiments about this on a pond on Clapham Common in 1770. Franklin observed that a drop covered about half an acre. Take a drop as $2\,\text{cm}^3$ and an acre as about $4 \times 10^3\,\text{m}^2$, how thick was the resulting film? Surprisingly, Franklin never made, or at least never mentioned, this calculation and it was about a century before it was done. ($2\,\text{cm}^3/2 \times 10^7\,\text{cm}^2 = 10^{-7}\,\text{cm}$.)

The human body has about $3000\,\text{in}^2 = 20.8\,\text{ft}^2 \simeq 1.93\,\text{m}^2$ of surface area. Atmospheric pressure is about 15 psi, so the total force on the body is $45{,}000\,\text{lb}$ or about $20{,}000\,\text{kg} = 20$ tons. Another source says the surface area is $20\,\text{ft}^2 \simeq 1.86\,\text{m}^2$. Using the value of atmospheric pressure found above, $10\,\text{tons}/\text{m}^2$, would give 19.3 tons or 18.6 tons.

Eddington estimated the number of particles in the universe is 10^{79}. The human brain contains about 10^{10} neurons [13].

The Great Pyramid of Cheops has volume about $2.7 \times 10^6\,\text{m}^3$ and mass about 7.2×10^6 tons. A year has about $3.16 \times 10^7\,\text{sec}$. The earth is about $1.5 \times 10^8\,\text{km}$ from the sun and Pluto is about 40 times as far. The seas contain about $1.4 \times 10^3\,\text{km}^3$ of water.

If one reads with a critical eye, one often finds numerical errors. Newspapers and popular books often have the following types of errors, due to innumeracy of editorial staff.

- The temperature may rise $5°\text{C}$ ($= 41°\text{F}$) by tomorrow.
- The creature was nearly $5\,\text{ft}$ ($15\,\text{m}$) tall.
- Use $3\,\text{m}$ screws to join the pieces.
- There are 212 degrees Fahrenheit in 100 degrees centigrade [2, pp. 152 & 284].

Bibliography

[1]   L. Russell Muirhead, ed. *Blue Guides: Denmark*. Benn, 1955, 3.

[2]   N. Blundell. The Trickster Quiz Book; Chancellor Press (Reed Consumer Books), 1994.

[3]   E. Boleman-Herring. *Insight Pocket Guide: Atlanta*. APA Publications, 1994, 14.

[4]   J. Clements. *Crazy — But True!* Armada, 1974, 15.

[5]   *Guide book to Leeds Castle*. 18.

[6]   C. G. Harper. *The City of London Guide*, 3rd ed. Burrow, 1930, 114.

[7] J. Henley. "Time finally hands over the clockmaker's £3m legacy." *The Guardian* (11 January 1999) 11.

[8] S. Leacock. *Happy Stories — Just to Laugh at.* 1945, 180–181.

[9] M. Leapman. The Companion Guide to New York. Collins & Prentice-Hall, 1983, 20 & 23.

[10] D. Lemon. *Everybody's Scrap Book of Curious Facts.* Saxon, 1890, 37.

[11] M. Lemon, ed. *The Jest Book: The Choicest Anecdotes and Sayings.* Macmillan, 1891.

[12] Michelin. *Dordogne Perigord Limousin Quercy* (Michelin Green Guide); 1st ed., Dickens Press, 1965, 181.

[13] L. Mottershead. *Investigations in Mathematics.* Blackwell, 1985, 32.

[14] J. Harker, ed. *Notes & Queries*, vol. 6. Fourth Estate, 1995, 232.

[15] *The Rough Guides: Paris.* The Rough Guide, 1997.

[16] Lilian and A. Russan. *Old London City: A Handbook.* Simpkin, Marshall, Hamilton, Kent & Co., 1924, 219.

[17] Sergio della Sala. "Mind Myths." in *The Fortean Times Book of UnConventional Wisdom.* John Brown Publishing, 1999, 21–23.

[18] C. Sutton and K. Markey. *More How Do They Do That?* Morrow, 1993, 92–94.

[19] L. Uday Lal. *Sura's Book of Amazing Fact.* Sura Books, Madras, no date, 37.

[20] J. Vinot. *Récréations Mathématiques.* Larousse & Boyer, 1860.

Chapter 13

Three Rabbits or Twelve Horses

The Three Rabbits image was first presented as a puzzle in the 19th century. It is a curious blend of a paradox and an optical illusion. The earliest example seen is in *The Girl's Own Book*, a popular 19th century American book. The first edition apparently was in 1831 and 42 editions have been traced in the US and Britain [1]. Here is a version from a facsimile of the 1833 edition.

13.1 The Three Rabbits Puzzle

The sources of this, posed as a puzzle, are found in English and American puzzle books from the 19th century. Paul Garcia at a British Society for the History of Mathematics meeting told me of an example in the stained glass of the church at Long Melford in Suffolk. Christopher Sansbury, then rector at Long Melford said the pattern was brought from Devon about 1600 and that it was fairly common on the eastern side of Dartmoor, mentioning Chagford, Widdecombe and North Bovey.[1]

The pattern was discovered at Paderborn and from there we learned about the existence of the pattern at Dunhuang and found images at the School of Oriental and African Studies in London.

---

[1]A version of this section appeared as Chapter 13: "Puzzles" in Tom Greeves, Sue Andrew & Chris Chapman, *The Three Hares: A Curiosity Worth Regarding*; Skerryvore Productions, South Molton, Devon, 2016; pp. 293–300.

Figure 13.1.   From [1].

By 2000, I was including the pattern in my talks on recreational mathematics, and from the audiences many further examples of the Three Rabbits, in Tibet, in the Islamic World, in illuminated manuscripts and in medieval Europe came to light.

However, there were very few references to the pattern as a puzzle. Indeed, there is hardly any written reference to the pattern. Inquiries in Dunhuang provided no further information, there are no other examples in China (except some in Tibet) and no reference to the pattern in Chinese literature.

Although there is no Buddhist trinity, the number three does enter into Buddhism in several ways. A Buddhist altar has to have three objects during worship: a reliquary representing the spiritual plane; a book representing the verbal plane and an image representing the physical plane. At Dunhuang, several caves have three versions of the Buddha — Past, Present and Future — and several writers have associated the three rabbits with past, present and future as a symbol of the eternal cycle of time. We do not know if there is any correlation between the presence of three Buddhas and of three rabbits.

The modern descriptions of Dunhuang show little awareness of the puzzle aspect [3]. In the Preface, we find the following. "What is particularly novel is the full-grown lotus flower painted in the center of the canopy design on the ceiling of Cave 407. In the middle of the flower there are three rabbits running one after the other in a circle. For the three rabbits only three ears are painted, each of them borrowing one ear from another. This is an ingenious conception of the master painter."

Further, there is no written reference to the pattern in the Islamic world, though there are many examples. One of the leading authorities on Islamic metalwork, Eva Baer, discusses the rabbits patterns in relation to Islamic art in general and mentions many of the examples. Much of this is given in her 1983 book, somewhat expanded in her *Islamic Ornament* (Edinburgh University Press, 1998). On pp. 118–124, she discusses the meaning of these images. She notes the antiquity of "animal wheel" images, mentioning Egypt, Skythians, Buddhist Central Asia, Islamic lands and medieval Europe. She says the rabbits pattern, with two, three or four hares, seems "to have enjoyed special favor among east Iranian artists" "between about the later 11th and the early 13th century." She discusses intertwined sphinxes and ducks. "It is much easier to trace the development of these motifs than to explain their Islamic meaning. ...the revolving [designs] remained a design of which no written records have come to my attention." By the 13th century, Islamic artists had lost interest in such patterns.

The puzzle aspect of the image goes back to about the 16th century in Europe. At Paderborn, Germany, the image is from about 1500 and there is a folk rhyme, but its date is not known:

> "Der Hasen und der Löffel drei,
> und doch hat jeder Hase zwei."

This translates as: "The hares and ears are three and yet each hare has two[, you see]."

There is a longer and more specific poem, "With United Powers and Ears", for which this is an English translation:

> Every hare has two ears.
> And here each has lost one.
> There should be six, there are only three.
> And appearance and being are different.
> How can the stone mason make an emblem?
> What thought is in the picture?
> The ears sit on the forehead,
> which flow into three heads.
> A third itself is here thus considered,
> which makes fear or joy.
> United, many things go easily even
> in the life of hares as in the life of men.
> And moreover this is, as you see,
> a playful image of the Trinity.

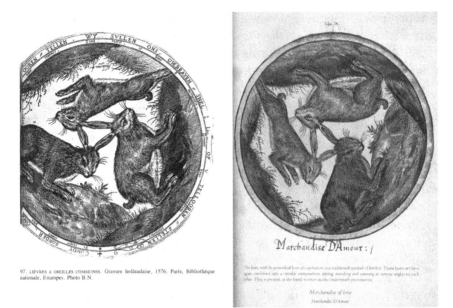

97. LIÈVRES À OREILLES COMMUNES. Gravure hollandaise, 1576. Paris, Bibliothèque nationale, Estampes. Photo B.N.

$\mathcal{M}$arcbandise $\mathcal{D}$Amour : /

The hare, with its proverbial love of copulation, is a traditional symbol of fertility. Three hares are here again combined into a circular composition, sitting, standing and running at various angles to each other. They represent, as the hand-written motto underneath pronounces:

Marchandise of love
Marchandise D'Amor

Figure 13.2.   From [2]. Also [4].

Unfortunately, neither of these Paderborn rhymes are dated. The earliest dated text for the three rabbits is a Dutch engraving of 1576. See Figure 13.2 (left) where parts of the Dutch and French texts have been cut off. Another version [4] also appears in Figure 13.2 (right), but the text has been painted over.

Fortunately, another version was discovered in 1997 in material stored in the Statens Museum for Kunst in Copenhagen, which is given below with my translation.

> Le Secret n'est pas grand lorsqun chacun le scait.
> Mais il est quelque chose a celui qui le faict.
> Tournez et retournez et nous tournerons aussi,
> Afin qu'a chacun de vous nous donnions du plaisir.
> Et lorsque nous aurez tournés faites compte de nos oreilles,
> C'est là que, sans rien déguiser, vous trouverez une merveille.

> The secret is not great when one knows it.
> But it is something to one who does it.
> Turn and turn again and we will also turn,
> So that we give pleasure to each of you.

Figure 13.3.   Use in advertisements.

> And when we have turned, count our ears,
> It is there, without any disguise, you will find a marvel.

Johann Heinrich Wilhelm Tischbein (1751–1829), best known for his portrait "Goethe in the Campagna", describes a decorated Easter egg he saw c. 1760: "On the egg are three hares with three ears, and yet each hare has its own two ears" [5].

Since about 1800, there are many descriptions of the pattern, and many of these point out the puzzling nature of the pattern. From about 1833, the puzzle began to appear in books of puzzles and in advertising cards in England and the USA. In the late 20th century, interesting variations began to be produced by artists.

James Dalgety produced a Millennium Puzzle Bench for Luppitt, Devon, parish church. This has an image of three pheasants sharing wings, designed by Julia Sparks, see Figure 13.6.

13.2   Four Horses, Twelve Horses and Other Puzzles

There are several other types of puzzling images based on anatomical rearrangement. We have recently learned of early 13th century examples of faces arranged in a circle and sharing eyes. These may derive from "tricephalic sculptures" where a head has three faces sharing eyes, so that the three faces have only four eyes all together, though only two eyes may be shown. Such images date back to the Gallo-Roman period in the 1st through 3rd centuries. These may derive from Roman images of the double-faced god Janus, which

PICTURE PUZZLES AND REBUSES

The Manx Rabbit Puzzle.—The only reason I can give for the name of this puzzle is that it was invented by a Manxman, that is a native of the Isle of Man. Now a rabbit is a rodent of the *genus Lepus* and those of the common variety are of the burrowing kind. As every college professor knows,

A-THE MANX RABBITS   B-THREE EARS FOR THREE RABBITS

FIG. 102.—THE MANX RABBITS PUZZLE

each rabbit is provided with a pair of beautiful, long silky ears, thus making six ears for three rabbits (how strange). The puzzle is to take the three rabbits shown at *A* in *Figure 102* and so place them that three ears will be enough to give all of them a pair apiece. It can be done as *B* clearly shows.

Figure 13.4.   A. Frederick Collins. *The Book of Puzzles*. D. Appleton and Co., NY, 1927. "The Manx rabbit puzzle", p. 153.

may derive from Hittite or Babylonian ideas, but this would take us too far from puzzles. In China, we noted that Avalokitesvara is often shown with multiple heads, with some sharing of features.

Another related and ancient variation of the idea is to have several bodies with one head or one body with several heads. The oldest, and perhaps the most beautiful, example has three fish sharing a head on a blue faience bowl, from 18th Dynasty Egypt, about 1500 BCE. This image has been popular ever since all around the world — there is a fine example in the Shaanxi History Museum, Xi'an, on an eared jug from the Yuan Dynasty (1271–1368).

Another famous example is from the Ajanta Caves in India, showing four deer sharing a head, done by Buddhist artists sometime from the 2nd century BCE to the 4th century CE.

Figure 13.5. Keith Kay; *Take a a Closer Look!* [sic]; Bright Interval Books (Brinbo), Bolton, Lancashire, 1988, p. 6. "The mysterious donkeys. Three donkeys with only 3 ears between them". Also Marjorie Newman. *The Christmas Puzzle Book.* Hippo (Scholastic Publications), London, 1990. "Kangaroos' ears", pp. 69 & 126.

Figure 13.6. Millennium Puzzle Bench for Luppitt, Devon, parish church.

A different type of puzzle based on anatomical rearrangement is Sam Loyd's Trick Mules of 1871. You have to cut the printed image into three pieces and put each rider on a mule. P. T. Barnum bought millions of copies of this for advertising. A nice version, is shown in Figure 13.9.

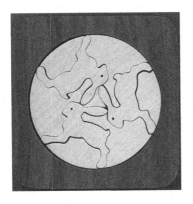
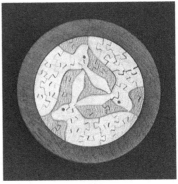

Figure 13.7.   Will Witham is making wooden "jigsaws" of the Three Rabbits pattern, both clockwise and anticlockwise; willw@grainwaves.co.uk and similar puzzles are being made by Decorative Puzzles (Mike & Gill Hayduk); mike@decorativepuzzles.co.uk.

Try to re-arrange the pictures in such a way that every hare has exactly two ears each.

The answer can be found somewhere in the exhibition!

Figure 13.8.   In 2004 Think Fun produced a web puzzle: Three Hares Matching Cards. "The object is to rearrange the six cards in such a way that every hare has exactly two ears". And another version was included in the handout leaflet for The Silk Road exhibition at The British Library in 2004.

In looking at 19th century puzzle books, it is clear that the Trick Mules derives from the "Dead Dogs Puzzle". The earliest known example is from 1849, see Figure 13.10. A similar image was popular throughout the world and long before the 19th century. In China it is known as "Sixi" or Four Happinesses. It is a general image of good luck and is often used on plates, toggles, scrolls, scroll weights and even on matchboxes, with the inscription: "Lian sheng gui zi tu" = "Many treasured sons diagram".

Figure 13.9.   Loyd's puzzle and solution. From: Gaston Tissandier, *Jeux et Jouets du jeune age*. G. Masson, Paris, n.d. [c. 1890], pp. 36–37.

The image is not restricted to China and Japan. It appears in India, Persia, Europe and Tibet. Perhaps the finest example is the following, with four heads but twelve horses. See Figure 13.11.

There are some old European examples. Baltrušaitis shows a nice Dutch engraving of 1576. A version from south Germany or Switzerland, late 15th century, had a rotating piece to allow the heads and bodies to be connected either way, and is one of the earliest known examples of a printed design with a movable part.

But the image goes back before printing. A version appears in the *Peterborough Psalter* of about 1310, now in Brussels. The earliest known example of the pattern in Europe is a sculpture on the Booksellers' Doorway of the Cathedral at Rouen, France, from 1275 to 1300.

In early 2004, Peter Rasmussen learned of several beautiful occurrences in Tibet, from the Kingdom of Guge, based at Tsaparong

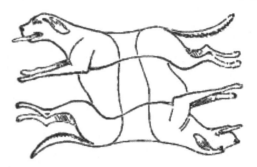

Figure 13.10.   From *Family Friend*, 1, 1849, 148 & 178.

Figure 13.11.   Persian? 17th century? Probably a leather cushion cover. Property of Martin Gardner, who inherited it from M. C. Escher. Jim Gardner says it was returned to Escher's heirs after Gardner's death.

(= Zhaburang) in the very southwest of modern Tibet. However, the dating of these is unclear — the Kingdom was active from the mid-9th century until about 1300 and somewhat later.

The description of the picture says this was a generally adopted composition in this time and place, see Figure 13.12. The book also says that they made "Auspicious revolving poems" which are a design

Figure 13.12. Two Auspicious Human Figures. From Dung-dkar Monastery in rTsa-mdav country (the area near Guge, Tibet).

of squares, each with a syllable so one can read it "in a crosswise, diagonal or circular way".

Other examples occur at other Guge sites. Since the dates of the Guge images are not known to any degree of precision, it is difficult to decide if they are earlier than the earliest European examples. So the question of the origin of this pattern remains open until we learn of more examples. In both Europe and Tibet, the earliest known examples are well developed, so one suspects that earlier examples must exist in or near both areas. It would seem natural for the image to have come from the Arabic world to both Europe and Guge, but no Arabic examples are yet known. It would be nice to discover some Arabic examples. No one at the 2004 Dunhuang conference could point to other early Chinese examples.

The Sixi image is an example of what visual psychologists call a "gestalt" phenomenon, where there are alternative views of the same image, like the classic duck/rabbit image. The two interpretations of the image, like the two interpretations of a Sixi image, compete in our mind. We can see one interpretation and then the other, but it is hard to see both at once. This is similar to Roger Penrose's famous Tribar illusion.

Bibliography

[1] L. M. Francis Child. *The Girl's Own Book*. Boston 1833.

[2] J. Baltrušaitis. *Le Moyen Age Fantastique Antiquités et exotismes dans l'art gothique*; Flammarion, Paris, 1981, p. 133.

[3] C. Shuhong and Li Chegxian. *The Flying Devis of Dunhuang*. China Travel and Tourism Press, Beijing, 1980, unpaginated.

[4] C.-P. Warncke, ed.; *Théâtre d'Amour*, Taschen, 2004, folio 78, figure 97; facsimile of an anonymous book of hand-colored engravings with MS annotations entitled *Emblemata amatoria* of 1620.

[5] "Wie der Hase zum Osterhase wurde," *Sächsische Zeitung*, 8 April 2004.

Printed in the United States
by Baker & Taylor Publisher Services